Internet Optical Infrastructure

János Tapolcai • Pin-Han Ho • Péter Babarczi
Lajos Rónyai

Internet Optical Infrastructure

Issues on Monitoring and Failure Restoration

János Tapolcai
Budapest University of Technology
 and Economics
Budapest, Hungary

Péter Babarczi
Budapest University of Technology
 and Economics
Budapest, Hungary

Pin-Han Ho
University of Waterloo
Waterloo, Canada

Lajos Rónyai
Institute for Computer Science and Control
Budapest University of Technology
 and Economics
Budapest, Hungary

ISBN 978-1-4614-7737-2   ISBN 978-1-4614-7738-9 (eBook)
DOI 10.1007/978-1-4614-7738-9
Springer New York Heidelberg Dordrecht London

Library of Congress Control Number: 2014945205

© Springer Science+Business Media New York 2015

This work is subject to copyright. All rights are reserved by the Publisher, whether the whole or part of the material is concerned, specifically the rights of translation, reprinting, reuse of illustrations, recitation, broadcasting, reproduction on microfilms or in any other physical way, and transmission or information storage and retrieval, electronic adaptation, computer software, or by similar or dissimilar methodology now known or hereafter developed. Exempted from this legal reservation are brief excerpts in connection with reviews or scholarly analysis or material supplied specifically for the purpose of being entered and executed on a computer system, for exclusive use by the purchaser of the work. Duplication of this publication or parts thereof is permitted only under the provisions of the Copyright Law of the Publisher's location, in its current version, and permission for use must always be obtained from Springer. Permissions for use may be obtained through RightsLink at the Copyright Clearance Center. Violations are liable to prosecution under the respective Copyright Law.

The use of general descriptive names, registered names, trademarks, service marks, etc. in this publication does not imply, even in the absence of a specific statement, that such names are exempt from the relevant protective laws and regulations and therefore free for general use.

While the advice and information in this book are believed to be true and accurate at the date of publication, neither the authors nor the editors nor the publisher can accept any legal responsibility for any errors or omissions that may be made. The publisher makes no warranty, express or implied, with respect to the material contained herein.

Printed on acid-free paper

Springer is part of Springer Science+Business Media (www.springer.com)

Foreword

With their seemingly infinite capacity to support the rapid transport of digital data to all corners of the globe, optical networks have without doubt been one of the mainstays of the modern Internet and the huge global economy that has grown up around it. As the line speed and capacity of optical transmission systems have grown to the levels of 10s of Terabits per second, then so too has the importance of being able to make networks of such high capacity line systems reliable and resilient. It goes without saying that an outage on such high transmission capacity trunks will affect very many end users of the network. Likewise, end-to-end network performance for the more specialist big data users, who will often call upon the network to support sustained high volume data transfers, using higher layer protocols, perhaps over long distances, can experience hugely impaired network performance where the underlying transmission is exhibiting high bit error rates or frequent protection/restoration events.

This means it is now of increasing importance that these high capacity optical networks exhibit as much survivability as they can, which is usually realised through the vehicles of service protection and restoration schemes in various combinations. In order to do this, such networks need robust control plane capabilities and these, in turn, rely on rapid and reliable detection of optical channel failure through robust monitoring techniques.

Subsequent to such failures, although the services should have survived with as minimal an outage as possible, it is essential that repairs are effected in an appropriately timely and cost-effective manner, otherwise multiple failures can rapidly diminish the survivability of the network. Especially where networks are spread over large geographical areas or are highly meshed (topologically complex) or the result of a federation of multiple autonomous operational domains then the ability to perform speedy and accurate fault localisation comes to the fore.

Contained in these pages the reader will find an engaging and detailed treatment of the topics of novel monitoring and failure localisation in all-optical networks, based on a canon of research work performed by the authors over a 10-year period.

The book starts with some conceptual introductions followed by an analytical approach in which the key problems are defined using formal methods, moving onto the description of new results addressing these problems and finally showing the verification of these results through simulation.

Chief Technology Officer Michael Enrico
DANTE and Technical Coordinator of GÉANT
  Open Call program
Cambridge, UK
2 April 2014

Preface

The book was originated from the authors' publications and doctoral dissertations in the area of optical network survivability and failure localization in the past decade, after going through extensive reorganization and editorial efforts. The intended readers of the book are those who are interested in gaining advanced knowledge and state-of-the-art research results in the topics. It serves as a good reference for graduate/training courses on network operations and carrier design. Therefore, we believe the book benefits students who are taking related graduate courses as well as network engineers for research of next-generation backbone networks. The interested reader can find demos of most of the methods discussed in the book at http://lendulet.tmit.bme.hu/demo/mtrail/.

The book is organized into three parts, each with specific objectives toward comprehension of the book scope. Part I is on fault management defined under *Generalized Multi-Protocol Label Switching (GMPLS)*. Part II talks about the use of *monitoring trails (m-trails)* for failure localization with a central controller for collecting the alarms. It defines the m-trail allocation problem by minimizing the number of m-trails, which relates to the complexity of the alarm collection and the size of the *alarm code table (ACT)*.

Part III studies the case where a node can perform failure localization based on the on–off status of the traversing m-trails. Such a node is called *Local Unambiguous Failure Localization (L-UFL)* capable node; and if all the nodes are L-UFL capable, the scenario is defined as *Network-wide L-UFL (NL-UFL)*. The scenario of L-UFL is further extended to an electronic signaling-free restoration framework where each node can automatically respond to the identified network failures and complete the required failure restoration task without waiting for any notification by another remote network entity. Such a framework is believed to achieve the ultimate goal of survivable optical network design: an ultra-fast restoration speed like in a ring network while enjoying optimal capacity efficiency as in a mesh topology.

Budapest, Hungary János Tapolcai
Waterloo, ON, Canada Pin-Han Ho
Budapest, Hungary Péter Babarczi
Budapest, Hungary Lajos Rónyai

Acknowledgments

We would like to thank first our families for their help and understanding throughout the book project. We would also thank our workplaces for giving us the time and support to conduct this project. János Tapolcai and Péter Babarczi are with the Department of Telecommunications and Media Informatics at the Budapest University of Technology and Economics, Hungary. Pin-Han Ho is working at the Department of Electrical and Computer Engineering, University of Waterloo, Canada. Lajos Rónyai is with the Computer and Automation Research Institute, Hungarian Academy of Sciences and with the Mathematical Institute of Budapest University of Technology and Economics.

The work of János Tapolcai and Péter Babarczi was partially supported by MTA-BME Lendület Future Internet Research Group and the High Speed Networks Laboratory (HSNLab) at the Budapest University of Technology and Economics and by the Hungarian Scientific Research Fund (OTKA grant K108947). Pin-Han Ho was supported by National Science and Engineering Research Council (NSERC), Canada. Péter Babarczi was supported by the János Bolyai Research Scholarship of the Hungarian Academy of Sciences (MTA). Lajos Rónyai was supported by the Hungarian Research Fund (OTKA grants NK105645, K77476) and TÁMOP-4.2.2/b-10/1-2010-0009. The work of the authors has been conducted with the financial assistance of the European Union under the FP7 GÉANT project grant agreement number 605243 as part of the MINERVA Open Call project.

Many thanks also to Bin Wu for introducing us to this field, and to our co-authors, in particular to Wei He, as well as to our colleagues Tibor Gyimóthy, Cecília Dudás, and Alija Pašić for their helpful comments during the preparation of the manuscript. Special thanks to the staff at Springer, especially Brett Kurzman and Rebecca Hytowitz, who were really helpful during the preparation of the book.

Contents

Part I Fault Management and Failure Restoration in Survivable Optical Networks

1 Introduction to Optical Fault Management 3
   1.1 Design Objectives of Survivable Routing Approaches 3
   1.2 Survivable Network Planning Framework: Notions and Notation 5
   1.3 Modeling Network Faults 7
       1.3.1 Root Causes of Network Faults 7
       1.3.2 The Shared Risk Link Group Failure Model 8
   1.4 GMPLS-Based Recovery in Transport Networks 9
   1.5 GMPLS-Based Fault Management and Device Configuration 12
   1.6 Summary and Outlook 13
   References 13

2 Failure Restoration Approaches 15
   2.1 Recovery Time Analysis 15
   2.2 Dedicated Protection 17
       2.2.1 1 + 1 Path Protection: A Widespread Protection Approach .. 17
       2.2.2 1 + 1 Realization Strategies for Better Resource Efficiency 19
   2.3 Shared Protection 20
       2.3.1 Pre-configured Protection (p-Cycles) 20
       2.3.2 Shared Backup Path Protection 22
       2.3.3 Shared Segment Protection 23
       2.3.4 Shared Link Protection 24
       2.3.5 Failure Dependent Protection 26
   2.4 Recovery Time Comparison of Protection Approaches 27
   2.5 Summary 28
   References 29

Part II Monitoring and Failure Localization in All-Optical Networks

3 Failure Localization Via a Central Controller 35
   3.1 Introduction 35
       3.1.1 Categorization of Optical Layer Failure
              Localization Schemes 37
       3.1.2 Problem Input 39
   3.2 UFL for Single Failures... 40
       3.2.1 Problem Definition 40
       3.2.2 Lower and Upper Bounds on the Number
              of (B)M-Trails 42
       3.2.3 An Optimal BM-Trail Solution in Densely Meshed Graphs.. 50
       3.2.4 An Optimal M-Trail Solution for Chocolate Bar Graphs 53
       3.2.5 An Essentially Optimal BM-Trail Solution
              for 2D Grid Topologies... 57
       3.2.6 Optimal BM-Trail Solution for Circulant graphs 61
       3.2.7 The RCA–RCS Heuristic Approach for UFL 64
   3.3 UFL for Multiple Failures 68
       3.3.1 Problem Definition and Background......................... 68
       3.3.2 Computational Complexity of UFL for Multiple Failures 71
       3.3.3 Optimal UFL Solution for Multiple Failures 73
       3.3.4 Sufficient and Necessary Conditions for SRLG UFL 79
       3.3.5 The Adjacent Link Failure Localization Heuristic
              Approach 85
       3.3.6 The LCC Heuristic Approach 89
       3.3.7 The CGT-GCS Heuristic Approach for M-Trail Allocation .. 94
   3.4 Performance Evaluation of UFL via a Central Controller............. 101
       3.4.1 Performance Evaluation of RCA–RCS for Single-Link
              UFL.. 101
       3.4.2 Performance Evaluation of AFL and LCC
              for Sparse-SRLG UFL... 105
       3.4.3 Performance Evaluation of CGT-GCS for
              Dense-SRLG UFL 108
   3.5 Summary 113
   References 114

4 Distributed Failure Localization 117
   4.1 Introduction 117
   4.2 Problem Definition 118
       4.2.1 Local Unambiguous Failure Localization 118
       4.2.2 An L-UFL Example 119
       4.2.3 State of the Art on L-UFL 119
       4.2.4 Network-Wide L-UFL 119
       4.2.5 An NL-UFL Example 121

	4.3	Bounds on Bandwidth Cost	121
	4.3.1	Lower Bound for General Graphs	122
	4.3.2	General Lower Bound for CGT	124
	4.3.3	Improved Lower Bound for Sparse Graphs	128
	4.3.4	Lower Bound for Dense Graphs	130
	4.3.5	Line Graphs	131
	4.3.6	Stars	132
	4.3.7	Complete Graphs	133
	4.3.8	Circulant Graphs	134
	4.4	The RSTA-GLS Heuristic Approach for NL-UFL	135
	4.4.1	Algorithm Description	135
	4.4.2	An Illustrative Example	138
	4.4.3	Performance Verification of RSTA-GLS	140
	4.5	Summary	146
	References		146

Part III An All-Optical Restoration Framework with M-Trails

5 Framework Introduction 151
5.1 Introduction 151
5.2 Signaling-Free Restoration Framework 152
  5.2.1 An Example on the Restoration Process 154
5.3 The Spare Capacity Allocation Problem 155
  5.3.1 The FDP-SCA Problem Formulation 155
  5.3.2 FDP Restoration Capacity Allocation 157
5.4 The Monitoring Resource Hidden Property 157
  5.4.1 Lower Bound on the Spare Capacity 158
  5.4.2 Dominance of Monitoring Resources 159
5.5 General Topologies with Multi-link SRLGs 160
5.6 Performance Evaluation 161
  5.6.1 Comparison of Signaling-Free Protection Methods 161
  5.6.2 Monitoring Resources Hidden 165
5.7 Summary 169
References 169

6 Global Neighborhood Failure Localization 171
6.1 Introduction 171
6.2 The G-NFL Scenario 172
  6.2.1 Introduction of G-NFL 172
  6.2.2 Resource Consumption by G-NFL 173
  6.2.3 Problem Definition 174
  6.2.4 Neighborhood 175
6.3 Bound Analysis 176
  6.3.1 Lower Bound for G-NFL 176

	6.4	G-NFL Heuristic	177
	6.5	Performance Evaluation	180
		6.5.1 Size of Neighborhood	180
		6.5.2 Restoration Time Analysis	180
		6.5.3 Coverlength of the G-NFL Solution with FDP	181
	6.6	Summary	184
		References	185
7	**Dynamic Survivable Routing with M-Trails**		187
	7.1	Spare Capacity Allocation in Dedicated and Shared Protection	187
		7.1.1 Suurballe's Algorithm	187
		7.1.2 Shared Protection	189
	7.2	Dynamic Joint Design Heuristic (DJH)	192
		7.2.1 Procedure of $MtrR$	194
		7.2.2 Generating M-Trails by $GenMtr$	196
	7.3	Summary	199
		References	200

Index 203

Part I
Fault Management and Failure Restoration in Survivable Optical Networks

The first part of the book provides an overview on the fault management and protection/restoration mechanisms defined under Generalized Multi-Protocol Label Switching (GMPLS), which serves as a background necessary for the subsequent study of the book. In particular, the book takes the GMPLS based recovery mechanism as a cross-layer operation between the IP and transport (or optical) layers, where the IP layer standard signaling protocols are extended to support the spare capacity reservation, fault management mechanism, and optical layer restoration.

Chapter 1 aims to provide a whole picture on the GMPLS based recovery framework, and position the task of optical layer monitoring and failure localization that will be in the focus of this book. Chapter 2 addresses the topic of survivable routing and spare capacity allocation by summarizing previously reported approaches to this purpose. As a basis for subsequent developments, this chapter discusses the causes of network faults and the possible shared risk link group (SRLG) models. Chapter 2 concludes with the general design principles for a survivable routing scheme which will be under the tradeoff between capacity efficiency and restoration complexity.

Chapter 1
Introduction to Optical Fault Management

Abstract In this chapter we introduce the survivable network design framework and identify the design goals of survivable network planning (i.e., resource efficiency and fault management complexity), and the trade-off between these objectives is discussed in protection and restoration approaches. A short summary on network faults and the shared risk link group model is presented, followed by a discussion of the phases of GMPLS-based fault recovery, and at the end of the chapter a brief summary is given.

1.1 Design Objectives of Survivable Routing Approaches

The main challenge in survivable optical network design is the wide range of possible failure events against which the deployed connections need to be resilient in order to fulfill the Quality of Service (QoS) requirements declared in the Service Level Agreement (SLA). Among others, the main QoS parameters include low delay, high availability, and minimal maximum outage time after a fault occurs. The last one has utmost importance from the perspective of upper layer protocols, owing to the fact that short disruptions in the optical domain are completely seamless for the applications. Thus, short outages won't degrade neither the performance nor user experience. In order to fulfill this short outage time requirement, providers can choose between pro-active approaches, i.e., *protection*, and re-active methods, i.e., *restoration*. Protection approaches pre-plan the recovery process of a *working lightpath (W-LP)* and reserve these resources in the network in advance, called *protection lightpaths (P-LP)*. The one-to-one correspondence between the W-LPs and P-LPs is referred to as *dedicated protection* (see Sect. 2.2 for details). In this approach, user data can be sent parallel along the W-LP and all corresponding P-LPs, thus, the maximum outage time after a fault occurs (referred to as *recovery time* of the protection approach) is minimal, quasi "instantaneous."

On the other hand, from a service provider's point of view, resource efficiency is also an important issue, as more users can be admitted if the key resources are managed carefully. Furthermore, network failures are rare events, thus, the P-LPs are intended to be used only for a short amount of time. Hence, it is a widely accepted strategy to share the allocated spare capacity among multiple P-LPs which aren't assumed to be activated at the same time. This is also referred to as *shared protection* (see Sect. 2.3 for details). In contrast to dedicated protection, a shared protection scheme relies on a suite of after-failure real-time mechanisms to restore the failed W-LPs, including failure localization, failure notification and failure correlation (*fault management*), and path selection and device configuration (*fault restoration*). Note that the detailed description of these after-failure tasks in Generalized Multi-Protocol Label Switching (GMPLS) is discussed in Sect. 1.4.

In the last decade, much research focused on the design of dedicated and shared protection schemes. Owing to its high complexity, several shared protection approaches were investigated, including shared backup path protection (SBPP), shared segment protection, shared link protection, failure dependent protection (FDP) (sometimes referred to as path restoration), and pre-configured protection [2, 7, 8, 17, 20, 22]. Although these approaches are built on basically different design goals, there are two objectives which are widely considered in their design when they are claimed in the context of all-optical mesh networks. These design goals are *resource efficiency* and *fault management complexity*. The former concerns the amount of *spare capacity* consumed, which is the capacity in terms of wavelength channels (WLs) reserved (but not necessarily configured) for the P-LPs; while the latter is measured as *recovery time*, which equals to the time period from the instant that the traffic is unexpectedly interrupted to that the traffic is completely restored.

In the selection process of the restoration or protection approach for a given provider network, the trade-off between resource efficiency and fault management complexity have to be considered. In the spectrum of protection scheme design, there exists a compromise between these two design goals, in which *FDP* and *pre-configured protection (dedicated protection and p-cycle)* are the two extremes. With FDP, each connection is assigned with multiple P-LPs, and one is activated for the restoration purpose according to the identified failure event. With the application of *stub release* (i.e., reusing the working capacity on the operational links of the disrupted connection for restoration), FDP has been recognized as having the optimal capacity efficiency and widely taken as the performance benchmark in the design of shared protection schemes. On the other hand, the FDP restoration process is subject to the highest control and signaling complexity that possibly yields the longest recovery time, mostly because the switching node has to precisely localize the failure event for real-time selection of a P-LP.

The other extremes are the pre-configured schemes that require minimum signaling, such as dedicated $1 + 1$ protection and p-cycles. Since the only after-failure real-time action is the reconfiguration of the two nodes responsible for switching and merging the affected working traffic, they require a minimum fault management complexity, thus, they can achieve very fast restoration. In contrast to any other counterparts, the pre-configured schemes can minimize the fault management complexity and possibly be implemented completely in the optical domain. Such simplicity and fast restoration speed are nonetheless at the expense of an excessive amount of spare capacity, i.e., poor resource efficiency.

The main goal of this book is to introduce a fault management framework with the application of *supervisory lightpaths (S-LPs)* (in Part III), which eliminates the need of such trade-off between the two design goals, i.e., provides a resource efficient protection approach with low fault management complexity.

1.2 Survivable Network Planning Framework: Notions and Notation

A *communication network* is a collection of directed links connecting transmitters and receivers. It may be represented by a directed graph $G = (V, E)$ with a vertex set V and an arc set E. There is a capacity of each link e denoted as $b(e)$, which represents the number of available WLs for routing the W-LPs, P-LPs, and S-LPs. The cost of using a unit bandwidth along link e is $c(e)$, which corresponds to the physical length of the link, or with applying traffic engineering the criticality of that link, etc.

We want to allow communication between selected nodes in the network, called source node s and destination node d of the connection, with a corresponding bandwidth demand b. The current trends point into the direction that in future networks the communication demands will arrive independently one after the other, without the knowledge of future incoming requests. This problem is often referred to as *dynamic routing*, where each $s-d$ demand has to be routed independently of other demands, considering the current state of the communication network $G = (V, E)$. Thus, the goal of dynamic routing is to establish a flow between s and d with flow value b. On the other hand, in dynamic *survivable routing* the goal is to deploy the W-LP and P-LP(s) in a way that after a failure of an arbitrary link set E' flow value $\geq b$ is maintained in the *failure graph* $G_{E'} = (V, E \setminus E')$. Note that protecting the connection against each link failure set would lead to $2^{|E|}$ different failure graphs, rendering the routing problem intractable. However, most network faults, especially multiple faults are rare events, thus, considering them in the route selection process is unnecessary. Thus, the considered failure events have to be selected carefully based on previous observations and operational premises, discussed in Sect. 1.3.

Although communication networks can efficiently be modeled by a graph $G = (V, E)$, there are a number of properties of real-world networks which restrict the set of graph models we need to consider in survivable routing. Shortly, we summarize the most common properties of these networks, and discuss graph generation processes, which produce real-world like networks for testing the protection and monitoring approaches.

Owing to the fact that optical cables are buried in common tunnels and manholes with other utilities (e.g., water pipes), the resulting topology is often considered as a planar network. Furthermore, as digging is a really expensive operation, burying cables between all optical cross connects is not affordable. Thus, the nodal degree of these topologies is around 2–5, and generally corresponds to "sparse graphs" in graph theory. Considering these observations, besides real network topologies protection and monitoring approaches are tested on planar 2-connected network topologies with different sizes and densities. For this purpose, the lgf_gen random graph generator of LEMON [12] was used, which randomly generates realistic planar 2-connected networks. The networks are classified according to the size of the largest facet of the planar graph, denoted by g. Clearly, a smaller value of g yields a more densely meshed topology. See also Fig. 1.1 for example networks.

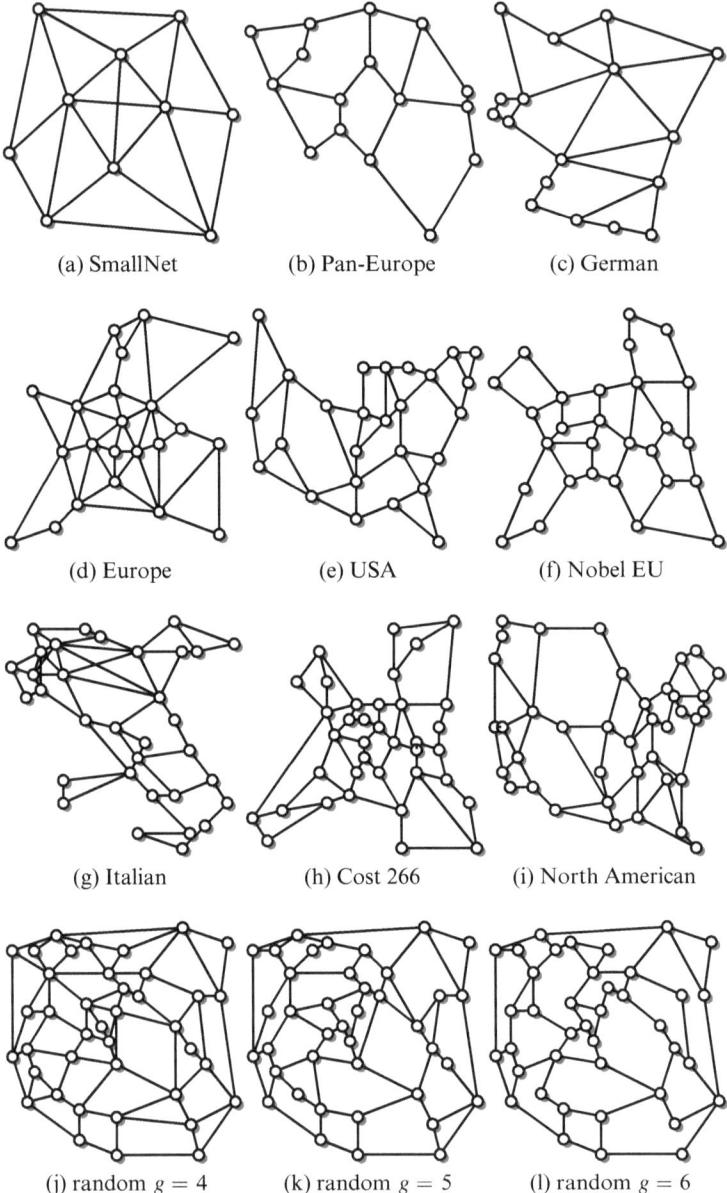

Fig. 1.1 Examples of backbone network topologies

1.3 Modeling Network Faults

At the graph representation level, the layered structure of the WDM optical network, the topology layout (e.g., physical location of the cables, common risks for multiple fibers) is lost. However, an exact model of possible failure events has utmost importance in survivable network design. Protecting only a small subset of failures could have high failure management complexity (owing to the real-time signaling to restore the connection), while selecting all possible failures as the input of the survivable routing problem results really poor resource efficiency. Thus, the service provider needs to select key failures carefully based on historical failure measurements or device parameters against which a connection needs to be resilient in order to fulfill the QoS requirement declared in the SLA. These most probable failure scenarios (Sect. 1.3.1) and their most common failure model (Sect. 1.3.2) are discussed in the rest of this section.

1.3.1 Root Causes of Network Faults

First, we list generic root causes of optical transport network failures following the studies [5, 6, 10, 11], where faults are categorized into the following six types:

Hardware Faults are reported to be 10–20 % of the network faults. The most frequently failed parts are the components with moving parts, such as cooling fans, hard disks, power supply.
   The complex integrated circuits may have built-in bugs, which may also cause problems in the operation of the equipment under very specific and rare circumstances. They are most likely filtered during the quality check.
Software Faults are rare in backbone networks, and much more common at higher layers. They are caused by design errors or faulty implementation.
Operator Errors are the human-type of errors or misconfiguration made during the design process, the normal operation of the network, the planned and unplanned maintenance. They are reported to be 35–50 % of the network faults.
Malicious Usage Attacks can be a physical intrusion, damage and/or robbery of equipment. It also can be any security attack as an illegal login, virus or Denial-of-Service (DoS) attacks to overload links and routers by transmitting an excessive amount of seemingly legitimate traffic to a certain destination. They are still rare in the fault reports of backbone networks.
Unexpected User Behavior are sudden and unexpected high user activity and load shifts that the network operator could not possibly foresee may be the cause of temporary performance degradation. It was less than 5 % in the reports.
Environmental Faults are 30–40 % of the network faults according to the reports. They are typically cable cut, power outage, extreme weather, disasters, fire, lightning, earthquake, flooding, tsunami, or terrorist attack.

Based on these observations, we conclude that node failures (the result of hardware faults and operator errors) and link failures (owing to environmental causes) are the two most probable failure scenarios of key network faults. However, faults caused, e.g., by extreme weather like Hurricane Sandy [16] in 2012 result in correlated outages. Thus, a precise fault model for dependent failure events has utmost importance in fault recovery of all-optical connections.

1.3.2 The Shared Risk Link Group Failure Model

The concept of *Shared Risk Link Group (SRLG)* was introduced to model failure dependencies between key network failures. An SRLG is defined as a group of network elements that could be links, nodes, physical devices, software or protocol identities, or a mix of them, etc., subject to a common risk of single failure, i.e., the SRLG expresses statistical dependencies between their failures. In practical cases an SRLG may contain multiple seemingly unrelated and arbitrarily selected links/nodes, and a link may belong to over 100 SRLGs [18]. For instance, two links belong to the same SRLG if they share the same tunnel or conduit, i.e., they are in the same *physical hierarchy*, which is related to the geographical topology of the network. On the other hand, considering the lightpaths (or IP/MPLS routes) built on top of this physical topology, two logical links belong to the same SRLG because they are deployed on the same physical link, i.e., they belong to the same *logical hierarchy*, which is related to the mapping of the logical/optical topology to the physical network [14].

For a connection demand between source node s and destination node d the SRLGs can be categorized as follows:

Protectable SRLG: An SRLG belongs to this type if the network still remains $s - d$ connected after the failure occurs, that is, the connection can be restored. In other words, the failed elements in the SRLG do not form a cut in the network topology; in this case, the working path affected by the failure is restorable.

Cut SRLG: An SRLG belongs to this type if the source and destination nodes are in different isolated fragments when the network is attacked by a failure. In other words, the failed elements in the SRLG form a cut between s and d. Thus, the interruption upon the corresponding working paths can never be restored. The cut SRLGs cannot be protected with any survivable routing method, thus, we assume that we are dealing with protectable SRLGs in the protection approaches.

We define that a *W-LP is involved in an SRLG* if it traverses through any network element that belongs to the SRLG. Two working paths share the same risk of a single failure if they are involved in any common SRLG. A W-LP is said to be SRLG-disjoint with its P-LP if the two paths are not involved in any common SRLG. In contrast to single-link failure resilience, when a general definition of the SRLGs is desired, a more complicated description and further elaborations are required to achieve an efficient implementation of any survivable routing algorithm for dedicated and shared protection. The SRLG-disjointness for a W-LP and P-LP path-pair is the major part of achieving 100 % restorability for the working data flow under the single failure scenario. However, if an arbitrary set of links can belong to the same SRLG, then the problem of finding an SRLG-disjoint path-pair between a pair of nodes in the network is NP-complete [4].

With $1+1$ dedicated protection, each W-LP and P-LP path-pair is pre-configured via the GMPLS signaling protocols, and is launched with the same copy of data transmitted between the source–destination $(s-d)$ pair during the normal operation. The two paths are SRLG-disjoint such that none of the single failures would affect both paths at the same moment. The 1:1 dedicated protection, on the other hand, only has the W-LP to be launched with data traffic while the capacity reserved by the P-LP is not in use. The best-effort traffic or any other signaling protocols (e.g., S-LPs in the monitoring framework) can make use of the spare capacity while the network is under normal operation. As an unexpected failure interrupts a W-LP (such as a fiber-cut or loss of signal (LOS)), the switching fabrics in the nodes along the corresponding P-LPs switch over to recover the original service supported by the W-LP.

For shared protection, the spare capacity taken by P-LPs can possibly be shared by some other P-LPs in order to improve capacity efficiency of the protection scheme. The P-LPs of different W-LP can reserve the same spare capacity if the W-LPs are not involved in any common SRLG. In other words, for shared protection the SRLG-disjointness must exist not only between the W-LP and its P-LP path-pair, but also among the W-LPs for which the corresponding P-LPs share spare capacity. The two types of SRLG-disjointness are also referred to as *SRLG constraint* for shared protection. It is clear that the implementation of shared protection imposes more disjointness requirements than that for dedicated protection. This leads to a fact that the development of shared protection schemes is generally more complicated.

1.4 GMPLS-Based Recovery in Transport Networks

After the design objectives of survivable routing and the possible failure events are discussed, the detailed steps of real-time tasks after a failure occurred are enumerated in the GMPLS environment. GMPLS has served as a building block of Internet backbone control and management and supports automatic fault restoration mechanisms in the network optical domain via a suite of signaling protocols (referred to as *GMPLS-based recovery*). These signaling protocols are extended from the original IP layer protocols such as Resource Reservation Protocol (RSVP) and Open Shortest Path First (OSPF).

GMPLS-based recovery defines the following five recovery phases [13, 15], which are defined as a standard sequence of generic operations performed when an optical layer failure event occurs. All the phases ($P_1 - P_5$) rely on electronic signaling via cross-layer protocol operations. In general, the detection of the failure event in the transport network will trigger the control plane for subsequent actions by way of GMPLS signaling protocol stacks. On the basis of [13], the following paragraphs explain each phase for the sake of completeness.

Phase 1 corresponds to the detection of the failure, which is a topology specific property, and independent of the applied recovery mechanism.

Failure Detection (P_1): Transport failure can be due to various reasons such as fiber-cut and component failure, which causes the detection of *loss of light (LOL)* or *LOS*. Failure detection strongly depends on the employed optical layer technology. For the detection of LOL, an optical cross connect (OXC) or a reconfigurable optical add-drop multiplexer (ROADM) can monitor LOL of each wavelength channel along a fiber (i.e., lambda monitoring) via a photodiode attached to each channel of an OXC. Such monitoring photodiodes have been norm to the commercial OXCs for the purpose of automatic power leveling, which is necessary when erbium doped fiber amplifier (EDFA) is employed.

The detection of LOS is usually performed with optical–electronic–optical (O/E/O) conversion. However, some advanced technologies could be employed to monitor transparent optical channels (i.e., lightpaths) in terms of their signal quality. For example, it is possible to monitor optical channel continuity [21] through laser bias currents or the optically received or transmitted power levels, and monitoring optical channel quality [20] through error-detecting codes, sampling methods, spectral methods, and indirect methods. All the optical layer monitoring techniques are subject to pros and cons. Error-detecting codes are the best for estimating the signal bit error rates (BERs) but they require access to electrical signals and may take more hardware complexity. Sampling methods are the most accurate for monitoring signals at the optical level, but they are complicated and expensive, and cannot be used in every element in the network. Lastly, the spectral time averaging methods are simply inspecting optical signals by ignoring all the distortion aspects, and are thus very inaccurate.

Phases 2–4 correspond to the fault management complexity of the protection and restoration approaches. Phases 2–4 are also referred to as *fault management*, which concerns how the control plane acquires the failure event information.

Failure Localization (isolation, P_2): The phase of failure localization is initiated immediately right after the failure detection, by which the network controller (or the deciding node(s)) obtain the locations (or the identities) of the data plane entities that detected the path(s)/link(s) failure. Such information can help to form a picture on where the failed transport network entities are located in order to achieve finer grained failure recovery.

The failure localization operation may be carried out via a suite of electronic signaling mechanisms performed in the data plane, such as the Link Management Protocol (LMP) with the traffic engineering and optical extensions. If such a protocol is not in place, the failure localization operation could alternatively be performed by way of collaboration between the signaling protocols in the control plane level and the data plane level.

1.4 GMPLS-Based Recovery in Transport Networks

Failure Notification (P_3): In the failure notification process, the intermediate nodes (usually the downstream nodes of the failure, i.e., the nodes between the failure and the destination node along the working path) are informed of the unavailability of the corresponding service. This process can be supported by either the data plane or control plane. The failure notification would possibly trigger either a protection or a restoration action via the control plane signaling. In general, the failure notification mechanism is reliable and efficient due to the use of acknowledgement and a disjoint control channel for aggregated messages for possibly multiple lightpaths.

Failure Correlation (P_4): A failure event could be detected by more than one neighboring network entities, whereby multiple failure notifications are issued. Thus a remote controller and/or switching nodes may receive multiple notifications due to the failure event. Failure correlation is for the remote controller and switching nodes to process the received notifications and make a correct decision upon the failure event.

Phase 5 is for the recovery (*path selection* and *device configuration*) of the affected working traffic from the failure event. The complexity of the former task depends on whether protection or restoration is applied, while the latter task has to be performed in almost all cases.

Fault Restoration (P_5): The affected traffic should be restored as soon as possible in order to minimize the vicious impact on the data plane, and it is done in the fault restoration phase. Once the switching node(s) of the affected W-LP completes the failure correlation, a decision is made on what to do in the failure recovery phase.

To facilitate fault restoration, the *transport network* (also called *optical domain*) can either rely on a dynamic *path selection* process that finds the protection lightpath after the failure event is identified and localized, or be equipped with a pre-planned protection plan where spare capacity is allocated before the occurrence of the failure event. The former usually takes more computation efforts thus leading to longer restoration latency. Neither can guarantee complete restoration. The latter, on the other hand, pre-plans the spare capacity as one or a set of P-LPs, each possibly corresponding to one or a set of considered failure events, such that the recovery of the affected W-LP can be guaranteed with a shorter recovery time. The price paid is the additionally allocated spare capacity, which is also referred to as *redundancy*.

No matter whether there is pre-planned spare capacity, *device configuration* to form the necessary P-LP(s) is required, and its latency contributes to the overall failure recovery time. We will see in Sect. 1.5 how the device configuration and the other fault management tasks are positioned in supporting the fault restoration process.

1.5 GMPLS-Based Fault Management and Device Configuration

As we have seen, fault management is defined under GMPLS as a set of sequentially performed after-failure real-time tasks $P_2 - P_4$, including failure localization, notification, and failure correlation. The former two tasks (i.e., localization and notification) in the transport network are defined by a series of electronic signaling mechanisms defined under the GMPLS-based protocols. By applying LMP coupled with a resource reservation protocol (such as RSVP-TE), each downstream node of a failed W-LP subject to LOL should send an alarm to its upstream node (i.e., the next node along the working path towards the source node). After receiving the alarm, the upstream node checks the corresponding input port and forwards the alarm to further upstream if the node is also subject to LOL. Otherwise, the faulty link is determined in the downstream, and the upstream node initiates protection or restoration procedures.

Such a GMPLS-based fault management process is subject to some weaknesses:

- Firstly, alarms could be simultaneously issued by multiple downstream nodes, and the number of alarms is determined by the number of W-LPs traversing through the faulty link(s) and the length of the W-LPs. The numerous alarms could easily lead to an *alarm storm* in the control plane and bring risks of crashing the whole network.
- Secondly, the GMPLS fault management could be vulnerable to multi-link SRLG failure events, due to the fact that a node can only be aware of the faulty link which is in the upstream of any W-LP traversing the node, but does not know the status of a link that is not traversed by any W-LP that goes through the link. Therefore, when a multi-link SRLG fails, a node may only be able to identify the failure of a subset of the links in the SRLG, and thus may select a P-LP that is nonetheless subject to the failure, too.
- Thirdly, due to extensive electronic signaling mechanism and nodal processing, the GMPLS fault management may take *hundreds of milliseconds for failure localization and notification*, and the delay is added to the overall recovery time. Note that a slow restoration not only causes data loss but also imposes vicious impact on the upper network and transport layer protocols such as OSPF and TCP.

To make the GMPLS fault management process better fit in the transport network, *link-based monitoring* [1, 3, 9, 19] has been considered such that basically every link is exclusively monitored via a single-hop S-LP. Once a failure occurs, the monitors subject to LOL issue alarm to the network controller or the corresponding decision nodes (e.g., edge routers) for the subsequent restoration process. Although being an effective solution for improving the GMPLS-based approach, the link-based monitoring approach strongly relies on control plane signaling for failure notification, which leads to *considerable control complexity and long recovery time*.

It also requires the same number of transmitters as the number of network links to support such a monitoring plane.

Several solutions based on various assumptions and design premises were reported in the past decade, and will be addressed in the fault management framework discussed in Part III of this book.

1.6 Summary and Outlook

In this chapter we identified the design goals of survivable network routing, namely resource efficiency and low fault management complexity, and discussed the trade-off between these objectives in protection and restoration approaches. After the introduction of the survivable network model a short summary on network failures and the SRLG model were provided, followed by the discussion on the phases of GMPLS-based failure recovery. Chapter 2 discusses the fault restoration approaches both from a capacity efficiency and from a recovery time point of view. In Part II of the book the issues on an optical monitoring framework based on S-LP design are introduced. Finally, in Part III the previously discussed results on monitoring and fault restoration are brought together into a common optical fault management framework, which eliminates the trade-off and satisfactorily addresses both design objectives of the Internet optical infrastructure.

References

1. Andersen R, Chung F, Sen A, Xue G (2004) On disjoint path pairs with wavelength continuity constraint in wdm networks. In: Proceedings of the twenty-third annual joint conference of the IEEE computer and communications societies (INFOCOM). IEEE, Hong Kong, China
2. Ellinas G, Stern TE (1996) Automatic protection switching for link failures in optical networks with bi-directional links. In: Proceedings of the global telecommunications conference (GLOBECOM), communications: the key to global prosperity, vol 1. IEEE, London, UK, pp 152–156
3. Ellinas G, Hailemariam AG, Stern TE (2000) Protection cycles in mesh wdm networks. IEEE J Sel Areas Commun 18(10):1924–1937
4. Ellinas G, Bouillet E, Ramamurthy R, Labourdette JF, Chaudhuri S, Bala K (2003) Routing and restoration architectures in mesh optical networks. Opt Netw Mag 4(1):91–106
5. Enriquez P, Brown A, Patterson DA (2002) Lessons from the pstn for dependable computing. In: Proceedings of the workshop on self-healing, adaptive and self-managed systems, vol 141
6. Gray J (1990) A census of tandem system availability between 1985 and 1990. IEEE Trans Reliab 39(4):409–418
7. Grover WD, Stamatelakis D (1998) Cycle-oriented distributed preconfiguration: ring-like speed with mesh-like capacity for self-planning network restoration. In: Proceedings of the IEEE international conference on communications (ICC), vol 1. IEEE, Atlanta, Georgia, USA, pp 537–543

8. Herzberg M, Bye SJ (1994) An optimal spare-capacity assignment model for survivable networks with hop limits. In: Proceedings of the global telecommunications conference (GLOBECOM). Communications: the global bridge, vol 3. IEEE, San Francisco, California, USA, pp 1601–1606
9. Hu JQ (2003) Diverse routing in optical mesh networks. IEEE Trans Commun 51(3):489–494
10. Kuhn DR (1997) Sources of failure in the public switched telephone network. Computer 30(4):31–36
11. Labovitz C (1999) Some observations on network failures. Tech. rep., NANOG 15. http://www.nanog.org/meetings/nanog15/presentations/Internet-Failures.ppt
12. LEMON (2010) A C++ library for efficient modeling and optimization in networks. http://lemon.cs.elte.hu
13. Mannie E, Papadimitriou D (eds) (2006) Recovery (protection and restoration) terminology for generalized multi-protocol label switching (GMPLS). RFC 4427
14. Papadimitriou D (2001) Inference of shared risk link groups. Internet Draft. http://tools.ietf.org/html/draft-many-inference-srlg-02
15. Papadimitriou D, Mannie E (eds) (2006) Analysis of generalized multi-protocol label switching (GMPLS)-based recovery mechanisms (including protection and restoration). RFC 4428
16. Quan L, Heidemann J, Pradkin Y (2013) Trinocular: understanding internet reliability through adaptive probing. In: Proceedings of the ACM SIGCOMM 2013 conference on SIGCOMM, SIGCOMM '13, pp 255–266
17. Stamatelakis D, Grover WD (2000) Network restorability design using pre-configured trees, cycles, and mixtures of pattern types. TRLabs, Edmonton. Tech. Rep. TR-1999-05
18. Strand J, Chiu A, Tkach R (2001) Issues for routing in the optical layer. IEEE Commun Mag 39(2):81–87
19. Suurballe JW, Tarjan RE (1984) A quick method for finding shortest pairs of disjoint paths. Networks 14(2):325–336
20. Thomassen C (1997) On the complexity of finding a minimum cycle cover of a graph. SIAM J Comput 26(3):675–677
21. Wasem OJ (1991) An algorithm for designing rings for survivable fiber networks. IEEE Trans Reliab 40(4):428–432
22. Wasem OJ (1991) Optimal topologies for survivable fiber optic networks using sonet self-healing rings. In: Proceedings of the global telecommunications conference (GLOBECOM), countdown to the new millennium. Featuring a mini-theme on: personal communications services. IEEE, Phoenix, Arizona, USA pp 2032–2038

Chapter 2
Failure Restoration Approaches

Abstract This chapter is on enumerating dynamic survivable routing schemes in mesh optical networks. By taking dynamic connection requests that arrive one after the other without any knowledge of future arrivals, the survivable routing schemes are required to allocate a disjoint working and protection path-pair for each connection request according to the current link-state. Without loss of generality, a working or a protection path is taken as a lightpath, either with a single wavelength if it is in a WDM network, or with some bandwidth allocated if it is in a spectrum-sliced elastic optical network supported by the optical orthogonal frequency division multiplexed technology.

2.1 Recovery Time Analysis

As we discussed in Chap. 1, restoration time is an important performance measure in a restoration scheme, which concerns the data loss and service discontinuity during a failure event. Here, we give a more detailed analysis of the requirements relevant for a protection or restoration scheme.

The best situation is that an optical domain failure event can be completely *hidden* from the upper layer control plane. For example, it is desired in some scenarios where a fiber-cut event can be restored purely in the optical domain while leaving the upper layer protocols (e.g., OSPF and TCP) unaware of, or at least with minimum impact due to, the event. To achieve this goal, *restoration time of a few tens of milliseconds is required*. However, the restoration process for a disrupted W-LP is composed of a number of real-time tasks such as fault management ($P_2 - P_4$) including failure localization, failure notification, failure correlation, and failure restoration (P_5), including path selection, and device configuration, as defined in the GMPLS-based recovery (see Sect. 1.4). Note that other components may also add to the total restoration time, such as that for failure detection (Phase 1) and traffic resumption. However, these are network specific properties and are independent of the applied protection approach. Thus, we won't consider them in the restoration time comparison of different protection structures. The restoration time (t_r) of a given survivable routing approach is defined as follows:

$$t_r = t_l + t_n + t_c + t_p + t_d, \qquad (2.1)$$

t_l is for failure localization (P_2), which is defined as the time between the instant that the failure is detected at a node, and the instant that the node reports the failure event to the decision node of the W-LP. Note that there could be multiple nodes adjacent to the failure which are to localize the failure.

t_n is defined as the time between the instant that the failure event is localized at some nodes, and the instant that the decision node receives the notification (P_3). t_n can be significantly reduced if the failure localization is performed at a node close to the node that performs traffic switchover. For example, a link restoration scheme (Sect. 2.3.4) localizes a failure exactly at the switching node, which should yield a negligible t_n. On the other hand, a path protection scheme usually takes much longer notification time due to multi-hop signaling in control plane via a network layer protocol.

t_c is the time between the instant that the decision node of the W-LP receives the failure notification and that it completes the failure correlation (P_4). This item is necessary since multiple notifications will be there and failure propagation may happen, where the decision node has to spend some time to wait for all possible notifications necessary to identify the failed SRLG.

t_p is the time for path selection ($P_5 - A$) at the decision node if multiple P-LPs are pre-planned (e.g., failure dependent protection (FDP) in Sect. 2.3.5). It is considered negligible, provided we have a precise identification of the failed SRLG.

t_d is defined as the time for setting up the P-LP ($P_5 - B$). The typical time to configure the switching matrix of an OXC is 10–20 ms. By using the GMPLS path setup message, device configuration is performed sequentially at every node along the P-LP, thus leading to a *latency of tens to hundreds of milliseconds* depending on the length and hop counts of the P-LP.

An example is given in Fig. 2.1 for a general restoration process. The W-LP and P-LP goes through $s - v_1 - v_2 - v_3 - d$ and $v_1 - v_4 - v_5 - v_3$, respectively. When a failure hits link (v_2, v_3), both nodes v_2 and v_3 localize the failure by taking t_l, and both of them send notifications to the corresponding decision node, i.e., node v_1 by taking t_n time. Note that the notification may be sent to s instead of v_1 for decision. No matter which node is the decision node, it needs to wait for t_c time to make sure only nodes v_2 and v_3 sent failure notifications. Then the decision node selects the protection path (t_p) for the identified failure and initiates a lightpath setup process

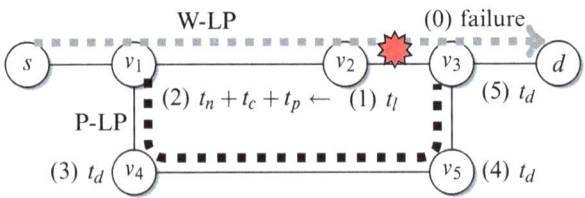

Fig. 2.1 An example of conventional restoration process

for device configuration (t_d) at nodes v_1, v_4, v_5, and v_3. After being acknowledged, the working traffic is switched at node v_1 and merges back to the W-LP at node v_3 to complete the restoration.

Obviously, dynamic path selection, i.e., finding the protection lightpath after the failure event (often referred to as *dynamic restoration* in GMPLS) would lead to a really slow restoration time owing to the fact that t_p could grow arbitrarily as the network size grows. Furthermore, the restorability of the connection cannot be guaranteed, which results a violation of the SLA at several points for a high QoS. Thus, we need to pre-plan the protection and allocate spare capacities for P-LPs before the failure occurs. There are several possible realization strategies of protection in the literature depending on the resource efficiency and fault management complexity trade-off. In the next sections, we discuss the two extreme approaches (1 + 1 and FDP) and several other methods in between.

2.2 Dedicated Protection

In dedicated protection there is a one-to-one correspondence between the working capacities and the protection capacities, i.e., a unit of spare capacity can be used only by a single connection in the restoration phase. Thus, dedicated protection approaches trade resource efficiency (about 80–120 % redundancy) for fault management complexity ($t_n = t_c = 0$ ms, $t_d = 20$ ms). However, there are several desirable properties (e.g., low operational complexity, low computational complexity, etc.) which make dedicated protection approaches favorable in a lot of application environments. Thus, we discuss shortly the dedicated protection methods from the simplest one (1 + 1) to the complex ones (network coded protection). We always keep in mind that we are interested in the most capacity efficient method with a given fault management complexity. If the network is upgraded, and higher management complexity can be handled by the network, the protection method can be adapted to the novel complexity level.

2.2.1 1 + 1 *Path Protection: A Widespread Protection Approach*

Inherent in the restoration mechanisms of self-healing rings [9, 10, 15, 17, 37, 40, 43, 44], the dedicated protection (i.e., 1 + 1 or 1:1) provides a very fast restoration service in mesh networks as well at the expense that the ratio of redundancy (i.e., the ratio of capacity taken by P-LPs compared to the W-LPs in the network) goes over 100 %. To implement dedicated protection in mesh optical networks for a single connection request, the physical routes for the W-LP and P-LP must be determined and configured before connection setup.

From a resource efficiency point of view 1:1 provides a better compromise than 1+1 protection. Although in both scenarios the protection paths are calculated in the same way as discussed in the next paragraphs, the operation of the two protection approaches differs. In 1:1 the spare capacity along the protection path can be used to send low priority traffic in the operational state of the network. After a failure occurs, the traffic on the P-LP can be pre-empted and used to protect the disrupted W-LP. As the data path has to be completely switched from the W-LP to the P-LP, a double-ended switching is required (both s and d have to switch from the W-LP to the P-LP). Thus, in the fault management phase, t_l, t_n, and t_c are nonzero values which lead to a slower recovery. On the other hand, in $1 + 1$ protection on the W-LP and P-LP the two copies of the same data are sent in parallel. Thus, after a failure occurs, only d has to perform protection switching based on the degraded signal quality along the W-LP. Thus, the fault management complexity highly decreases as real-time signaling is completely avoided. This property makes $1+1$ *the fastest and simplest protection approach* for the price of 100 % redundancy. Both 1+1 and 1:1 dedicated protections are very suitable for those platinum services supported by the lambda-switching transmission mode for an ultra-fast restoration. For those connections requiring 100 % restorability that can tolerate some extent of data loss and delay caused by any failure, shared protection becomes a better choice by trading the restoration time for resource efficiency.

It has been reported that the multi-link SRLG-disjoint diverse routing problem for dedicated protection is NP-complete [2, 11, 23], while solving the least-cost link- or node-disjoint working and protection path-pair in directed graph can be done by Suurballe's algorithm [39] with polynomial computational complexity (see Sect. 7.1.1 for details). Besides finding two disjoint paths with minimal total cost, more complex variants of the disjoint path problem have been studied as well. In [50] the problem of finding a disjoint path-pair such that the length of the shorter is minimized was addressed, while in [26] the length of the longer path was minimized. However, in the realm of optical protection total minimal cost is the desired property of the disjoint path-pair. Some previous studies have focused on dedicated protection in mesh WDM networks with the wavelength continuity constraint, where the W-LP and P-LP may take different wavelengths. In [23], a proof of NP-completeness is given to the diverse routing problem considering the wavelength continuity constraint, in which the logical topology of the network is given as part of the input. This implies that the lightpaths have already been set up between specific pairs of nodes, and the diverse routing effort turns out to be the selection of a path-pair for a given connection request. In [2], an NP-completeness proof is given to the same problem yet with different assumptions from those which are considered in [23]. The case where each link can carry only a certain subset of wavelengths is considered, and the proposed algorithm attempts to establish a disjoint pair of W-LP and P-LP to satisfy the wavelength continuity constraint.

2.2.2 1 + 1 Realization Strategies for Better Resource Efficiency

As we have discussed in Sect. 1.3.2, in the SRLG model, it is not enough that the two paths of 1 + 1 protection are node disjoint (and therefore link-disjoint): an SRLG failure (i.e., simultaneous failure of multiple links) could lead to the disruption of both paths, if both the W-LP and the P-LP of the same connection are involved in that SRLG. Furthermore, in some scenarios it is worth to relax that we seek for a good solution in the form of an SRLG-disjoint path-pair. Thus, the desired properties of 1 + 1 are generalized into the *1+1 path protection (1PP) functionality* for SRLGs, which is defined as the ability to tolerate a single SRLG failure in either the respective W-LP or P-LP(s), while the W-LP and all P-LP(s) are calculated and fully signaled in advance (pre-configured switching matrices in OXCs, $t_d = 0$ at intermediate nodes), and data being sent during the whole lifetime of the connection (spare resources are in hot-stand-by, thus, $t_l = t_n = t_c = 0$ ms) [4].

We call the methods fulfilling all of these criteria as *protection approaches with instantaneous recovery* (well below 50 ms, see Table 2.2 for details), as these approaches have the lowest recovery time and best fault management complexity, as neither packet retransmission nor flow rerouting required after a failure occurred. Thus, 1PP dedicated protection methods completely eliminate the need of any real-time signaling in the control plane ($t_n = t_c = 0$ ms). Furthermore, applying one of the following two realization strategies, besides minimal fault management complexity, resource efficiency could be kept on a tolerable level as well.

Diversity Coding 1 + 1 Path Protection: this protection approach operates by establishing three or more disjoint paths between s and d [30]. Different data packets are transmitted on each path (e.g., d_1 on W-LP$_1$ and d_2 on W-LP$_2$); in addition, a redundancy packet is formed by simply XORing the data packets together and sent on the P-LP of the connection ($d_1 \oplus d_2$). 1PP is ensured as information from only two out of three paths is needed for the reconstruction of the lost packet. Note that more complex coding schemes may also be used.

General Dedicated Protection with Network Coding: the concept of network coding (NC) allows for in-network modification of data by coding several data packets together inside the network and decoding them at a proper receiving node [1]. Mostly known for its favorable multicast throughput characteristics, NC is also a suitable tool for path protection [5]. General Dedicated Protection with Network Coding (GDP-NC) [3] was proposed to solve the resource efficiency issue of 1 + 1 with a general protection structure rather than a disjoint path-pair. The design goal in GDP-NC is to find a minimal cost arbitrary protection structure to survive any SRLG failure given as the input of the problem. Thus, it generalizes all previous 1PP dedicated protection methods (including 1 + 1), while it provides *optimal resource efficiency* among them.

Although GDP-NC has optimal resource efficiency among 1PP approaches, we face the trade-off that we have to pay the price of a more complex operation, as the intermediate nodes have to be able not only to forward but also to code incoming data. However, 1PP is still maintained, i.e., the fault management complexity is still very low in comparison with shared protection approaches, discussed in Sect. 2.3.

We note here that Suurballe's algorithm provides a node-disjoint (and therefore link-disjoint) solution for $1+1$ and diversity coding 1PP in the single-link and node failure case in polynomial-time, while a linear program exists for GDP-NC [3]. However, when considering SRLGs, all 1PP approaches become NP-complete (for $1+1$ and diversity coding 1PP it is shown in [11], while for GDP-NC in [3]).

2.3 Shared Protection

It has been observed that the resource sharing between different P-LPs can substantially reduce the redundancy required to achieve 100 % restorability at the expense of longer recovery time. To perform survivable routing for shared protection, the spare capacity along each link can be divided into two classes: either *sharable* or *non-sharable*, according to the location of the W-LP, which, in turn, gives rise to the *SRLG constraint* for shared protection when the P-LP is determined. In the next sections we enumerate the main shared protection families, following the line of the resource efficiency and fault management complexity trade-off, starting from the pre-configured protections and heading towards the most resource efficient FDP approach. We note that several realization strategies exist for each protection method; however, we only give some insight to the design issues in each family, while providing references to the interested reader.

2.3.1 Pre-configured Protection (p-Cycles)

Pre-configured protection falls in the category of shared protection because each piece of reserved spare capacity can be used by multiple W-LPs that will not be commonly affected by a considered single failure event. Different from all other shared protection schemes where spare capacity is reserved but not configured, a pre-configured protection scheme has spare capacity allocated in a specific shape and pre-configured so as to achieve ultra-fast and electronic signaling-free failure restoration. It is considered the only version of shared protection that can be completely implemented in the optical layer without the aid of any upper layer control plane signaling, thanks to preconfigured intermediate nodes along P-LPs, which achieves a very fast restoration process due to the fact that the only after-failure real-time action is the reconfiguration of the two nodes responsible for switching and merging the affected working traffic. The switching and merging nodes can localize and correlate the failure via in-band monitoring of the W-LP that is assumed to be bidirectional without waiting for any further failure notification

2.3 Shared Protection

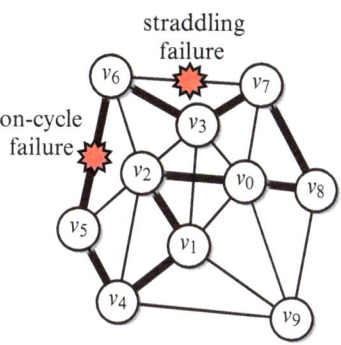

Fig. 2.2 P-cycle restoration: (1) An on-cycle failure (e.g., $v_6 \to v_5$) is restored by using the rest of the cycle. (2) A straddling failure (e.g., $v_6 \to v_7$) can be restored by using both directions of the cycle

($t_n = 0$). Such simplicity and fast restoration speed are nonetheless at the expense of *consuming the highest redundancy among shared protection approaches*.

Pre-configured cycle (or p-cycle) is one of the most developed proposals for performing pre-configured protection [12, 16, 41]. An example is shown in Fig. 2.2, where the only cycle contained in the network is ($v_0 \to v_2 \to v_1 \to v_4 \to v_5 \to v_6 \to v_3 \to v_7 \to v_8 \to v_0$) with an on-cycle failure on link $v_5 \to v_6$ (or $v_6 \to v_5$). P-cycle allows working and spare capacity to be different from link to link, in which the "straddling" failure is defined (e.g., link $v_6 \to v_7$ in Fig. 2.2). The link on which the working capacity can be subject to a straddling failure must have both of the end nodes traversed by the corresponding cycle so as to improve the capacity sharing. It is notable that a p-cycle is bidirectional; in other words, a *p-cycle in the network topology may yield two directed cycles* in the design space with opposite orientations.

One of the best advantages of using p-cycle is that it can be implemented in mesh networks composed of rings, each of which works as a self-healing ring unit. However, it is clear that the resource sharing is only between the spare capacities provided by a directed cycle. Therefore, the resource efficiency is worse than all the other shared protection schemes. In addition, due to a fixed shape of pre-configured spare capacity, it is inevitable to yield excess spare capacity along some links. However, it requires very high computational time to achieve a true optimal design.

With the knowledge of all working capacities in the network, optimal p-cycle allocation can be formulated as an Integer Linear Program (ILP), in which the number of copies for each directed cycle (i.e., any predefined cycle) with a basic bandwidth unit is determined to achieve a certain design criterion (e.g., 100 % restorability for all working capacity in the network). The number of copies for each different directed cycle pattern can be 0 or any positive integer. One widely employed approach for solving the p-cycle allocation problem is via cycle enumeration, where a set of candidate cycles is prepared, and the goal of the optimization problem is to minimize cost due to the consumed spare capacity allocated for each copy of the directed cycles.

On the other hand, the widely adapted ILP formulation for p-cycle with candidate cycle enumeration has many disadvantages. Since the size of the candidate set soars exponentially as the network size increases, cycle enumeration is far from scalable, and a large candidate set certainly incurs a huge number of variables in the ILP. This slows down the optimization process (if not making it intractable). To reduce the size of the candidate set, some heuristic cycle pre-selection algorithms [8, 14, 27, 52] can be used, or the length of candidate cycles can be limited. As a result, only a subset of all cycles is chosen for ILP optimization. Obviously, this will adversely affect the quality of the solution. An approach was proposed in [29] to generate p-cycle solutions using only fundamental cycles, defined as cycles without straddling links. This approach needs a preprocessing step to enumerate all fundamental cycles and construct the straddling link information. Besides, it cannot efficiently handle p-cycle design in non-planar networks. Hence, some other studies turned to a p-cycle design without cycle enumeration [36, 45], and claimed to yield true optimal solutions via solving ILP models. A flow-based ILP was formulated in [36], and a much faster version of ILP using cycle-exclusion was given in [45].

2.3.2 Shared Backup Path Protection

For dedicated path protection $(1 + 1)$, the derivation of the least-cost W-LP and P-LP path-pair can be solved easily by Suurballe's algorithm. However, this is not the case for shared protection. More generally, an ILP is introduced for solving the shared backup path protection (SBPP) problem in a single step (see Sect. 7.1.2.2). To overcome the relatively large complexity implied by the NP-completeness of ILP, a number of heuristics for this purpose have been proposed.

Most studies on the routing of the working and protection path-pair for shared protection are based on the technique of deriving the least-cost P-LP given the corresponding W-LP. A general method is called Two-Step-Approach that can be applied to both shared and dedicated protection. However, the Two-Step-Approach could block the connection request when no P-LP can be derived for the given W-LP, although with a different W-LP an appropriate P-LP could be selected. Thus, a one step approach using Suurballe's algorithm is more favorable for dedicated protection. On the other hand, for shared protection Two-Step-Approach is a good candidate for survivable routing (see Sect. 7.1.2.1 for more details).

There are numerous papers on the Two-Step-Approach. In [6, 18, 21, 22, 35, 38, 40, 46, 49, 51] k-shortest paths are inspected iteratively between each $s - d$ pair one after the other until the least-cost working and protection path-pair is derived; while the studies in [6, 22, 48] investigate how to estimate the location of the working path such that the corresponding protection path can take the most advantage of spare capacity resource sharing. There are schemes as well for jointly deriving a working and protection path-pair, called Iterative Two-Step-Approach (ITSA) [22], Potential Backup Cost (PBC) [32], and Maximum Likelihood Relaxation (MLR) [22].

2.3.3 Shared Segment Protection

Although SBPP could reduce the resource consumption of $1 + 1$, owing to the fact that the P-LP could only be set up after a failure occurred, failure localization (t_l) and failure notification (t_n) could take excessive time, leading to high fault management complexity. This could be even slower in large networks with long W-LPs and P-LPs, where the node responsible to localize the failure could be far from the ones who require performing protection switching. In order to keep the failure information local to the failure (thus, reducing t_n), shared segment protection (SSP) is introduced.

With SBPP, the only nodes that can switch the affected W-LP are s and d. However, in SSP an alternative design is possible by allowing any intermediate node along a W-LP to switch and merge traffic for performing local restoration. In such a circumstance, a working path is segmented into multiple protection domains, each of which behaves as a self-healing unit performing local restoration once the corresponding working segment is subject to a failure. SSP has been reported to provide a larger extent of resource sharing to achieve a better performance than the SBPP schemes by relaxing the SRLG constraint, as well as a guaranteed restoration service by manipulating the length of the working and protection path segments.

The following is an example to show the capacity saving in using SSP rather than SBPP. In Fig. 2.3, the network contains a working path W-LP$_1$: $s \to v_1 \to v_2 \to v_3 \to v_4 \to v_5 \to d$ and its spare capacity along the physical route P-LP$_1$: $s \to v_6 \to v_2 \to v_7 \to v_4 \to v_8 \to d$. Let another working path W-LP$_2$: $v_2 \to v_3 \to v_4$ be allocated with its protection path P-LP$_2$: $v_2 \to v_7 \to v_4$. In the case of SBPP the spare capacity taken by P-LP$_1$ can never be shared by P-LP$_2$ since both of the working paths are involved in the same SRLG. For the SSP, on the other hand, W-LP$_1$ is segmented into multiple segments, each of which is assigned with a switching and merging node and protection path segment. P-LP$_2$ can then share the spare capacity taken by the protection path segments in the first and the third protection domain. In this case, the total amount of non-sharable spare capacity in the network for a specific working path segment is reduced, which yields better resource efficiency. It is notable that the working segments of adjacent protection domains can be designed to overlap such that node protection is also achieved.

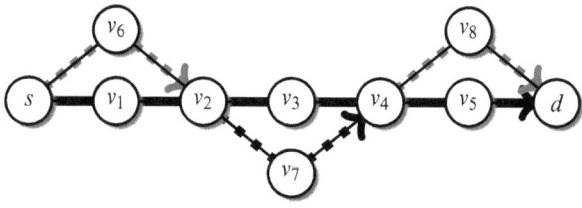

Fig. 2.3 SSP with node pairs $[s, v_2]$, $[v_2, v_4]$, and $[v_4, d]$ being the switching and merging nodes for the first, second, and third protection domain, respectively

Although better resource efficiency can be achieved, the fault management complexity is increased by using SSP compared with SBPP. The reasons are as follows: firstly, a larger computation complexity is taken due to the necessity to determine the switching and merging nodes for each protection domain and the derivation of the spare link-state for multiple working path segments. Secondly, a higher requirement on hardware responsiveness and signaling efforts is imposed in the recovery process.

Work on SSP is briefly surveyed as follows. The approach developed in [35] is characterized by separating the tasks of routing and spare capacity allocation into two subsequent processes, in which the spare capacity sharing for each link is not considered until the physical routes of the backup segments are defined. In [38] and [25], two similar dynamic algorithms are proposed to implement shared link protection (SLP), in which each link is switched to the backup path from its immediate upstream node and merge back to the original path at its immediate downstream node and at any of the downstream nodes, respectively. In [19, 21], a framework called Short Leap Shared Protection (SLSP) along with a dynamic survivable routing scheme called Cascaded Diverse Routing (CDR) is proposed to realize SSP, in which the enumeration of k-shortest paths in each segment of the working path segment is performed. Basically, with CDR the switching and merging nodes along the working path are predefined such that the working and protection path segments in each protection domain are selected using the same approach as where shared path-based protection is performed. Therefore, the performance can be improved if the location of the switching and merging nodes can be included into the design space. The study in [20] attempts to overcome this limitation by enumerating the simple cycles with a length limitation which intersect with the given working path and containing exactly a single switching and merging node-pair at least at one hop distance from each other. These cycles are called candidate cycles for the W-LP. The protection domain allocation problem is formulated into a shortest path searching problem after sorting and labeling each candidate cycle.

2.3.4 Shared Link Protection

Opposite to SBPP, shared link (span) protection is the other extreme case of SSP, where every protection domain of a W-LP is a single-link. Thus, different from SBPP and SSP, a survivable routing scheme can be designed such that the working capacity of each link is protected with one or multiple P-LPs corresponding to each considered failure event on the link.

To implement SLP, a simple approach is to prepare a set of candidate routes for each link, and the candidate routes can be obtained offline by using a k-shortest path searching algorithm. The bandwidth required along each candidate route can be calculated by solving an ILP. To impose a limit on the recovery time, an upper bound in terms of physical distance (or without loss of generality, hops) is imposed upon the candidate protection routes of each link. Then, the algorithm assigns spare capacity for the working capacity of each link and determines the amount

2.3 Shared Protection

of bandwidth (or the number of lightpaths) in each restoration route that yields a minimum amount of total spare capacity. This scheme can guarantee 100% restorability for any single link failure upon the existing working capacity in the network in the device configuration phase.

An example of SLP is given in Fig. 2.4 and Table 2.1, where 8 W-LPs are launched in a network with 7 nodes and 10 links. The algorithm first finds all the eligible protection routes for every link that may be subjected to a failure. The protection routes should be up to a length limit 3 in hop count in this case. We have all the protection routes corresponding to each directed link listed in Table 2.1. Then, an optimization process is performed to derive the minimum amount of spare capacity needed to restore any possible failure along each link.

From the point of view of recovery time, SLP is better than SSP or SBPP, because it keeps the scope of the failure local ($t_n = 0$), as the endpoint of a link is responsible

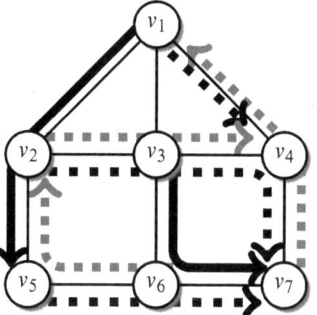

Fig. 2.4 Working lightpaths on a 7-node 10-link example network

Table 2.1 The eligible protection routes in Fig. 2.4 for each link with W-LP. The length limit is 3 hops

Link	Protection route	Link	Protection route
$v_1 \to v_2$	$v_1 \to v_4 \to v_3 \to v_2$ $v_1 \to v_3 \to v_2$	$v_2 \to v_1$	No W-LP to protect
$v_1 \to v_3$	No W-LP to protect	$v_3 \to v_1$	No W-LP to protect
$v_1 \to v_4$	$v_1 \to v_2 \to v_3 \to v_4$ $v_1 \to v_3 \to v_4$	$v_4 \to v_1$	$v_4 \to v_3 \to v_1$ $v_4 \to v_3 \to v_2 \to v_1$
$v_2 \to v_5$	$v_2 \to v_3 \to v_6 \to v_5$	$v_5 \to v_2$	$v_5 \to v_6 \to v_3 \to v_2$
$v_2 \to v_3$	$v_2 \to v_1 \to v_4 \to v_3$ $v_2 \to v_5 \to v_6 \to v_3$	$v_3 \to v_2$	No W-LP to protect
$v_3 \to v_4$	$v_3 \to v_2 \to v_1 \to v_4$ $v_3 \to v_6 \to v_7 \to v_4$	$v_4 \to v_3$	No W-LP to protect
$v_3 \to v_6$	$v_3 \to v_4 \to v_7 \to v_6$ $v_3 \to v_2 \to v_5 \to v_6$	$v_6 \to v_3$	No W-LP to protect
$v_4 \to v_7$	$v_4 \to v_3 \to v_6 \to v_7$	$v_7 \to v_4$	$v_7 \to v_6 \to v_3 \to v_4$
$v_5 \to v_6$	$v_5 \to v_2 \to v_3 \to v_6$	$v_6 \to v_5$	$v_6 \to v_3 \to v_2 \to v_5$
$v_6 \to v_7$	$v_6 \to v_3 \to v_4 \to v_7$	$v_7 \to v_6$	No W-LP to protect

2.3.5 Failure Dependent Protection

Finally, we discuss FDP, as the other extreme of pre-configured protection in the design spectrum of protection approaches. In FDP, each W-LP can be assigned with multiple P-LPs, and only one is activated for the restoration purpose, depending on the failed SRLG. FDP has long been recognized as a way to achieve optimal resource efficiency and widely taken as the performance benchmark in the design of shared protection schemes. However, the FDP restoration process is *subject to the highest control and signaling complexity* that possibly yields the longest recovery time, mostly because the switching node requires precise knowledge of the failed SRLG for real-time selection of a P-LP.

On the other hand, FDP [7, 12, 16, 24, 28, 33, 34, 41, 42, 47] was reported as the most general approach for spare capacity allocation that can possibly *achieve optimal resource efficiency among* **all protection methods**. Basically, an FDP scheme protects a W-LP with multiple end-to-end P-LPs, each corresponding to a specific failure event. The pre-planned P-LPs are typically only partially disjoint from the W-LP, while the W-LP is restored by activating one of the P-LPs at the source according to the identified failure event that unexpectedly interrupted the W-LP. The consumed capacity can be minimized via *stub release*, where the bandwidth originally used by the interrupted W-LPs is released for being used by the activated P-LPs. Note that the use of the released bandwidth of a W-LP is not limited to its P-LP, but possibly used by a P-LP of another unexpectedly interrupted W-LP.

Routing of P-LPs given a set of W-LPs for FDP has been extensively studied in the past decade. The method is formally called Spare Capacity Placement/Allocation for Path Restoration with stub release, and is generally called *path restoration*. Detailed descriptions can be found in [24, 28, 47], [13, Chap. 6], [31, Chap. 9.5.2] for both ILP formulations and heuristic approaches. To be specific, for FDP path restoration the ILP in [31] and Successive Survivable Routing (SSR) algorithm in [28] are two widespread approaches. The SSR algorithm iteratively routes link-disjoint P-LPs for a W-LP corresponding to each SRLG affecting the W-LP one at a time, using shortest path search based on the latest spare capacity information.

With the best flexibility, FDP with stub release generalize all the other shared protection flavors by removing all possible limitations in the selection of P-LPs for a W-LP, thus, leading to the optimal resource efficiency and usually taken as the benchmark in the performance evaluation among different shared protection schemes. Nonetheless, such a superb performance behavior is at the expense of the highest demand on the network failure state, which generally requires extensive electronic signaling support from the control plane (see Chap. 5 for examples on

FDP). For example, the source node of the interrupted W-LP should be able to precisely localize the failed link(s) before the corresponding P-LP can be selected and activated for restoration. With this regard, the FDP restoration process requires a suite of real-time signaling mechanisms for failure localization, failure notification, and failure correlation ($P_2 - P_4$). Along with the P-LP setup (or device configuration P_5), FDP is generally considered impractical in an all-optical network where failure recovery time of a W-LP should be within a few tens of milliseconds. Thus, a systematic approach, such as S-LPs (monitoring trails, m-trails) is needed to eliminate these real-time signaling efforts, and enables all-optical and signaling-free failure restoration under SRLG failures for FDP.

Part III of the book investigates this FDP-based restoration framework for efficient failure localization, and eliminates the need of resource efficiency and signaling complexity trade-off.

2.4 Recovery Time Comparison of Protection Approaches

Finally, we summarize the resource efficiency and fault management complexity of the previously discussed protection approaches in Table 2.2, which contains the candidates for all-optical failure recovery.

First, we mentioned that $1+1$ and 1PP dedicated protection methods completely eliminate the need of any real-time signaling in the control plane, thus, provide quasi instantaneous recovery for the price of 100% redundancy. In these approaches, only the destination node d needs to perform protection switching, which makes these approaches the fastest ones with about 20 ms recovery time. On the other hand, in 1:1 protection, after a failure occurs, the low-priority traffic needs to be pre-empted, while double-ended switching is required at both the source s and destination node d. Therefore, the end-nodes of the connection have to be notified of the failure ($t_n > 0$), which has to be localized previously ($t_l > 0$), resulting in an increased recovery time compared to $1+1$. However, the spare resources of 1:1 could be used to carry low priority traffic.

Table 2.2 Typical recovery times (in ms) and redundancy (i.e., percentage of spare capacity compared to working capacity)

Protection	Fault management			Failure restoration		Total	Redundancy (%)
	t_l	t_n	t_c	t_p	t_d	t_r	
$1+1$	0	0	0	0	20	20	100
1:1	10	10	10	0	20	50	100
1PP	0	0	0	0	20	20	80
p-Cycle	10	0	10	0	20	40	75
SBPP	10	20	20	0	50	100	60
FDP	10	30	30	10	50	130	40

Considering shared protection, p-cycles yield about 40 ms for fault management by having $t_l \approx 10$ ms for the failure to be localized at adjacent nodes, t_c as 10 ms mainly for failure correlation, and $t_d = 20$ ms is required at the two on-cycle nodes for switch fabric configuration. An important reason for such short recovery time by a pre-configured scheme is that *no multi-hop signaling is required*. Other shared protection schemes have to significantly rely on a suite of signaling mechanisms in their restoration processes. In SLP, t_l is similar to that of p-cycle, but establishing a P-LP requires additionally a few tens milliseconds for device configuration. For path restoration schemes such as SSP, the recovery time can be up to a hundred milliseconds due to the multi-hop signaling notification process and longer P-LPs. Further, the restoration process for FDP yields the longest recovery time due to the following three reasons.

1. Since precise failure correlation is needed at the decision node, $t_n + t_c$ is the longest possible.
2. Multiple P-LPs are in place for each W-LP, yielding nonzero t_p.
3. The P-LPs are not pre-configured, thus a regular P-LP setup process is required.

On the other hand, from a resource efficiency point of view these protection and restoration schemes perform completely different, as we have seen previously, and shown in the last column of Table 2.2.

It is desired to facilitate a general protection scheme to achieve all-optical restoration like pre-configured protection without sacrificing the resource efficiency like in FDP and SSP. In the rest of the book, a systematic approach by way of monitoring trails (m-trails) is investigated to enable all-optical and signaling-free failure restoration under SRLG failures. We note here, that with the application of S-LPs, as in the m-trail framework the failure recovery time of FDP can be reduced to about 50 ms, while the redundancy increases from 40 to 50 %.

2.5 Summary

In this chapter, we explored the dedicated and shared protection approaches from the perspective of resource efficiency and fault management complexity. We discussed general dedicated protection approaches (1PP), which have optimal resource efficiency among all dedicated protection methods, while fault management complexity is similar to the simplest $1+1$ method. On the other hand, we investigated shared protection approaches, and discussed that FDP has optimal resource efficiency among all protection methods. However, from the viewpoint of fault management complexity, it requires precise failure localization and a very complex signaling mechanism to restore the disrupted connection. In Part II of the book the theoretical background of unambiguous failure localization (UFL) will be discussed. UFL in the optical domain enables FDP to perform failure localization and completely avoid failure notification, which highly reduces its fault management complexity. This makes the protection framework introduced in Part III, applicable in practice.

References

1. Ahlswede R, Cai N, Li S, Yeung R (2000) Network information flow. IEEE Trans Inf Theory 46(4):1204–1216
2. Andersen R, Chung F, Sen A, Xue G (2004) On disjoint path pairs with wavelength continuity constraint in wdm networks. In: Proceedings of the twenty-third annual joint conference of the IEEE computer and communications societies (INFOCOM), vol 1. IEEE, Hong Kong, China
3. Babarczi P, Tapolcai J, Ho PH, Médard M (2012) Optimal dedicated protection approach to shared risk link group failures using network coding. In: Proceedings of the IEEE international conference on communications (ICC), pp 3051–3055
4. Babarczi P, Biczók G, Overby H, Tapolcai J, Soproni P (2013) Realization strategies of dedicated path protection: a bandwidth cost perspective. Elsevier Comput Netw 57(9):1974–1990
5. Belzner M, Haunstein H (2009) Performance of network coding in transport networks with traffic protection. In: Proceedings of the ITG symposium on photonic networks. VDE, Vienna, Austria, pp 1–7
6. Bouillet E, Labourdette JF, Ellinas G, Ramamurthy R, Chaudhuri S (2002) Stochastic approaches to compute shared mesh restored lightpaths in optical network architectures. In: Proceedings of the twenty-first annual joint conference of the IEEE computer and communications societies (INFOCOM), vol 2. IEEE, New York, USA, pp 801–807
7. Choi H, Subramaniam S, Choi H (2004) Loopback recovery from double-link failures in optical mesh networks. IEEE/ACM Trans Netw 12(6):1119–1130
8. Doucette J, He D, Grover WD, Yang O (2003) Algorithmic approaches for efficient enumeration of candidate p-cycles and capacitated p-cycle network design. In: Proceedings of the fourth international workshop on design of reliable communication networks (DRCN). IEEE, Banff, Alberta, Canada, pp 212–220
9. Ellinas G, Stern TE (1996) Automatic protection switching for link failures in optical networks with bi-directional links. In: Proceedings of the global telecommunications conference (GLOBECOM), communications: the key to global prosperity, vol 1. IEEE, London, UK, pp 152–156
10. Ellinas G, Hailemariam AG, Stern TE (2000) Protection cycles in mesh wdm networks. IEEE J Sel Areas Commun 18(10):1924–1937
11. Ellinas G, Bouillet E, Ramamurthy R, Labourdette JF, Chaudhuri S, Bala K (2003) Routing and restoration architectures in mesh optical networks. Opt Netw Mag 4(1):91–106
12. Frederick M, Datta P, Somani A (2006) Sub-graph routing: a generalized fault-tolerant strategy for link failures in wdm optical networks. Elsevier Comput Netw 50(2):181–199
13. Grover WD (2003) Mesh-based survivable networks: options and strategies for optical, MPLS, SONET and ATM networking. Prentice Hall PTR, Upper Saddle River
14. Grover WD, Doucette J (2002) Advances in optical network design with p-cycles: joint optimization and pre-selection of candidate p-cycles. In: Proceedings of the existing and emerging architecture and applications/dynamic enablers of next-generation optical communications systems/fast optical processing in optical transmission/VCSEL and all-optical networking. IEEE, IEEE-LEOS Summer Tropical Meetings, Mont Tremblant, Quebec, Canada, pp WA2–WA49
15. Grover WD, Stamatelakis D (1998) Cycle-oriented distributed preconfiguration: ring-like speed with mesh-like capacity for self-planning network restoration. In: Proceedings of the international conference on communications (ICC), vol 1. IEEE, Atlanta, Georgia, USA, pp 537–543
16. Grover W, Doucette J, Clouqueur M, Leung D, Stamatelakis D (2002) New options and insights for survivable transport networks. IEEE Commun Mag 40(1):34–41
17. Herzberg M, Bye SJ (1994) An optimal spare-capacity assignment model for survivable networks with hop limits. In: Proceedings of the global telecommunications conference (GLOBECOM), communications: the global bridge, vol 3. IEEE, San Francisco, California, USA, pp 1601–1606

18. Ho PH, Mouftah H (2001) Issues on diverse routing for wdm mesh networks with survivability. In: Proceedings of the tenth international conference on computer communications and networks (ICCCN). IEEE, Scottsdale, Arizona, USA, pp 61–66
19. Ho PH, Mouftah HT (2002) A framework for service-guaranteed shared protection in wdm mesh networks. IEEE Commun Mag 40(2):97–103
20. Ho PH, Mouftah HT (2004) A novel survivable routing algorithm for shared segment protection in mesh wdm networks with partial wavelength conversion. IEEE J Sel Areas Commun 22(8):1548–1560
21. Ho PH, Tapolcai J, Cinkler T (2004) Segment shared protection in mesh communications networks with bandwidth guaranteed tunnels. IEEE/ACM Trans Netw 12(6):1105–1118
22. Ho PH, Tapolcai J, Mouftah HT (2004) On achieving optimal survivable routing for shared protection in survivable next-generation internet. IEEE Trans Reliab 53(2):216–225
23. Hu JQ (2003) Diverse routing in optical mesh networks. IEEE Trans Commun 51(3):489–494
24. Iraschko RR, MacGregor M, Grover WD (1998) Optimal capacity placement for path restoration in STM or ATN mesh-survivable networks. IEEE/ACM Trans Netw 6(3):325–336
25. Kodialam M, Lakshman T (2001) Dynamic routing of locally restorable bandwidth guaranteed tunnels using aggregated link usage information. In: Proceedings of the twentieth annual joint conference of the IEEE computer and communications societies (INFOCOM), vol 1. IEEE, Anchorage, Alaska, USA, pp 376–385
26. Li CL, McCormick S, Simchi-Levi D (1990) The complexity of finding two disjoint paths with min-max objective function. Discrete Appl Math 26(1):105–115. http://dx.doi.org/10.1016/0166-218X(90)90024-7. http://www.sciencedirect.com/science/article/pii/0166218X90900247
27. Liu C, Ruan L (2004) Finding good candidate cycles for efficient p-cycle network design. In: Proceedings of the 13th international conference on computer communications and networks (ICCCN). IEEE, Chicago, IL, USA, pp 321–326
28. Liu Y, Tipper D, Siripongwutikorn P (2005) Approximating optimal spare capacity allocation by successive survivable routing. IEEE/ACM Trans Netw 13(1):198–211
29. Nguyen HN, Habibi D, Phung VQ, Lachowicz S, Lo K, Kang B (2006) Qrp02-5: joint optimization in capacity design of networks with p-cycle using the fundamental cycle set. In: Proceedings of the global telecommunications conference (GLOBECOM). IEEE, San Francisco, California, USA, pp 1–5
30. Øverby H (2006) Cost evaluations of the protected network layer packet redundancy scheme. J Opt Netw 5(10):747–763
31. Pióro M, Medhi D (2004) Routing, flow, and capacity design in communication and computer networks. The Morgan Kaufmann series in networking, Morgan Kaufmann Publishers/Elsevier
32. Qiao C, Yoo M (1999) Optical burst switching (obs)—a new paradigm for an optical internet^{1}. J High Speed Netw 8(1):69–84
33. Ramamurthy S, Mukherjee B (1999) Survivable wdm mesh networks. II. Restoration. In: Proceedings of the IEEE international conference on communications (ICC), vol 3. IEEE, Vancouver, British Columbia, Canada, pp 2023–2030
34. Ramasubramanian S, Harjani A (2006) Comparison of failure dependent protection strategies in optical networks. Photonic Netw Commun 12(2):195–210
35. Saradhi CV, Murthy CSR (2002) Dynamic establishment of segmented protection paths in single and multi-fiber wdm mesh networks. In: The convergence of information technologies and communications. International Society for Optics and Photonics, Boston, MA, USA, pp 211–222
36. Schupke D (2004) An ilp for optimal p-cycle selection without cycle enumeration. In: Proceedings of the eighth working conference on optical network design and modelling
37. Stamatelakis D, Grover WD (2000) Network restorability design using pre-configured trees, cycles, and mixtures of pattern types. TRLabs, Edmonton. Tech. Rep. TR-1999-05
38. Su X, Su CF (2001) An online distributed protection algorithm in wdm networks. In: Proceedings of the IEEE international conference on communications (ICC), vol 5. IEEE, Helsinki, Finland, pp 1571–1575

References

39. Suurballe JW, Tarjan RE (1984) A quick method for finding shortest pairs of disjoint paths. Networks 14(2):325–336
40. Thomassen C (1997) On the complexity of finding a minimum cycle cover of a graph. SIAM J Comput 26(3):675–677
41. Wang D, Li G (2008) Efficient distributed bandwidth management for mpls fast reroute. IEEE/ACM Trans Netw 16(2):486–495
42. Wang H, Modiano E, Médard M (2002) Partial path protection for wdm networks: end-to-end recovery using local failure information. In: Proceedings of the seventh international symposium on computers and communications (ISCC). IEEE, Taormina, Italy, pp 719–725
43. Wasem OJ (1991) An algorithm for designing rings for survivable fiber networks. IEEE Trans Reliab 40(4):428–432
44. Wasem OJ (1991) Optimal topologies for survivable fiber optic networks using sonet self-healing rings. In: Proceedings of the global telecommunications conference (GLOBECOM), countdown to the new millennium. Featuring a mini-theme on: personal communications services. IEEE, Phoenix, Arizona, USA, pp 2032–2038
45. Wu B, Yeung K, Ho PH (2010) Ilp formulations for p-cycle design without candidate cycle enumeration. IEEE/ACM Trans Netw 18(1):284–295
46. Xin C, Ye Y, Dixit S, Qiao C (2001) Joint lightpath routing approach in survivable optical networks. In: Proceedings of the Asia-Pacific optical and wireless communications conference and exhibit. International Society for Optics and Photonics, Beijing, China, pp 139–146
47. Xiong Y, Mason LG (1999) Restoration strategies and spare capacity requirements in self-healing atm networks. IEEE/ACM Trans Netw 7(1):98–110
48. Xu D, Qiao C, Xiong Y (2002) An ultra-fast shared path protection scheme-distributed partial information management, part ii. In: Proceedings of the 10th IEEE international conference on network protocols (ICNP). IEEE, Paris, France, pp 344–353
49. Xu D, Xiong Y, Qiao C (2003) Protection with multi-segments (promise) in networks with shared risk link groups (srlg). In: Proceedings of the annual Allerton conference on communication control and computing, 1998, vol 40. The University, IEEE/ACM Transaction on Networking, 11(2):248–258
50. Xu D, Chen Y, Xiong Y, Qiao C, He X (2006) On the complexity of and algorithms for finding the shortest path with a disjoint counterpart. IEEE/ACM Trans Netw 14(1):147–158
51. Yuan S, Jue JP et al (2002) Shared protection routing algorithm for optical network. Opt Netw Mag 3(3):32–39
52. Zhang H, Yang O (2002) Finding protection cycles in dwdm networks. In: International conference on network protocols IEEE international conference on communications (ICC), vol 5. IEEE, New York, USA, pp 2756–2760

Part II
Monitoring and Failure Localization in All-Optical Networks

Part II of the book presents the state-of-the-art of optical layer failure localization using multi-hop supervisory lightpaths, known as monitoring trails (m-trails). The on-off status of each m-trail is obtained and correlated for achieving a failure localization decision. We focus on the m-trail allocation problem for a set of m-trails according to the SRLGs to be considered in the network. The problem of m-trail allocation is formulated under two scenarios. One is with a central control for collecting the alarms issued by the monitors of those m-trails interrupted by the failure, where electronic signaling is required as a vehicle for relaying the alarms to the central controller. The other is that each node can locally sense the on-off status of the traversing m-trails, such that the node can individually make the failure localization decision. Obviously, the second scenario has the merit of not relying on the alarm notification like the first scenario; but this can be achieved by consuming more bandwidth (or wavelength links) and complexity for solving the m-trail allocation problem. Chapters 3 and 4 are devoted to the above two scenarios.

Chapter 3
Failure Localization Via a Central Controller

Abstract To achieve fast Unambiguous Failure Localization, an essential problem for the network operators is to determine how to efficiently probe the network elements such that the number of probes is the minimum. By launching a set of m-trails, the transmitter of each m-trail constantly probes the health of the links along the m-trail, and the monitor at the receiver issues an alarm once detecting any irregularity. A failure may interrupt multiple m-trails which incurs a set of alarms. The m-trails should be allocated such that the network controller can uniquely and precisely localize the failure state according to the issued alarms. The chapter is on the m-trail allocation problem by introducing algorithms and approaches in presence of single and multiple link failures, respectively. With single-link failures, an essentially optimal construction for m-trail allocation is provided for lattice topologies. For general topologies, a suite of heuristics are presented, including Random Code Swapping (RCA–RCS) for single-link failures, Adjacent Link Failure Localization and Link Code Construction for adjacent link failures, and Greedy Code Swapping (CGT-GCS) for dense-shared risk link group failures based on combinatorial group testing.

3.1 Introduction

In transparent optical networks, failure localization is a very complicated issue that has been extensively investigated [5, 20, 27, 32, 42, 44, 45, 49–52].

Due to the lack of optoelectronic regenerators the impact of a failure propagates without any electronic boundary and a single failure can trigger a large number of redundant alarms [24, 44]. With failure recovery protocols at different network layers, various failure management mechanisms with specific built-in failure management functionality could be adopted. Thus, a failure event at the optical layer (such as a fiber-cut) may also trigger alarms in other upper protocol layers [10], possibly causing an *alarm storm*. This not only increases the management cost of the control plane, but also makes failure localization difficult. Therefore, isolating failure recovery within the network optical domain is essential to solve the problem, which will be enabled by an intelligent and cost-effective failure monitoring and localization mechanism dedicated to the network optical layer. One of the most commonly adopted approaches is to deploy optical monitors responsible for generating alarms when a failure is detected. The alarm signals then flood in the control plane of the optical network such that any routing entity can localize the failure and perform traffic restoration in a timely manner. Minimizing the number of alarm signals while achieving Unambiguous Failure Localization (UFL) serves as the major target in the development of a failure localization scheme. In addition, reducing the bandwidth consumption for fault monitoring should also be considered.

In general, a link is a conduit of multiple fibers, and each fiber supports one or multiple wavelengths. Thus, it is intuitive to monitor a link cut event by monitoring a single wavelength along the link, and for this purpose a monitor is activated at one of the end nodes that will issue an alarm once a loss of light (LoL) is detected along the wavelength channel. This is also referred to as *link-based monitoring*, which requires $|E|$ active alarms (or monitors) to monitor each link independently, where $|E|$ is the number of network links. In this case, an alarm code with a length $|E|$ is required in order to identify any single-link cut event.

However, it is not considered an efficient approach to dedicatedly allocate a monitor for each link. To resolve this situation, the studies in [49–52] introduced the monitoring-cycle (m-cycle) concept, which is a pre-configured loop-back lightpath terminated by an optical power monitor and launched with supervisory optical signals. When any link along the cycle is cut, the supervisory lightpath is interrupted, and the failure will be detected by the optical power monitor, and the monitor will issue an alarm to the rest of the network.

To ease the limitation on the cycle constraint of the monitoring structure, [44] introduced the concept of Monitoring-Trail (m-trail), where the model is based on an enumeration-free Integer Linear Program (ILP) approach. M-trail is proved to yield much better performance by employing monitoring resources in the shape of trails— a monitoring structure that generalizes all the previously reported counterparts. However, due to the huge computation complexity in solving the ILP, only network topologies with small sizes (such as 30 nodes) can be handled. A similar monitoring structure called "permissible probes" was considered in [18]. The study focused on theoretical proofs and asymptotic bounds, while the strength and flexibility of using tree structures for launching probes was little explored in possible practical scenarios. More detailed comparisons and descriptions of the monitoring structures (e.g., cycles, trails, trees, etc.) can be found in [48].

In this chapter, we investigate the m-trail design problem for single-link and later for multi-link UFL, and aim at obtaining deeper understanding and insight into the problem. In particular, our focus is on the impact of topology diversity to the problem solutions. The chapter first provides a categorization of the current state of the art failure localization schemes. Next, it analytically derives the minimum lengths of alarm codes for several graph topologies, which is followed by m-trail allocation algorithms developed for general topologies. They are verified by extensive simulation on thousands of randomly generated topologies.

3.1.1 Categorization of Optical Layer Failure Localization Schemes

In an optical layer monitoring scheme, a link failure is detected and localized based on the on-off status of some lightpaths. These schemes can be categorized according to the following three aspects:

- the type of failures they can identify,
- the constraints on the lightpaths used for network status acquisition, and
- the failure localization time.

3.1.1.1 The Types of Failures

A failure could be either *hard* or *soft* [26]. A hard failure involves immediate interruption due to link and/or node function disorder typically due to fiber cuts or network node failure, while a soft failure simply degrades the performance of one or multiple wavelength channels. In this book we deal with hard failures only. The failures can be further categorized according to their geographic locations. Most previous studies focused on *single-link failures*, which nonetheless account for just some fraction of total failures. Node failure events also contribute to the total failures. The rest of the failure events, including operational errors, power outage, and denial of service (DOS) attack, etc., could hit multiple links/nodes in the network simultaneously. When modeling these failures, often a list of *Shared Risk Link Groups* (SRLGs) is defined by the network operators. An SRLG is a group of network elements subject to a risk of simultaneous failure (see Sect. 1.3.2). We call it *sparse SRLG* scenario if the number of SRLGs in the network is similar to the number of links. In this case each SRLG may consists of single or multiple links and they typically correspond to a failure at a specific geographic location [7, 8]. For example, two links over the same river share the common risk of failure because of flooding. If every single and double-link failures are part of the list of SRLGs, we call it *multiple failure (dense SRLG)* scenario. In this case there is a large number of highly overlapped and densely distributed SRLGs.

3.1.1.2 The Constraints on the Monitoring Lightpaths

Network elements can be monitored either via *in-band* or *out-of-band* monitoring [17]. The former obtains the network failure status only by monitoring the existing (or operational) lightpaths, while the latter launches supervisory lightpaths for failure status acquisition. Out-of-band monitoring is favored for its simplicity and data independence, even at the expense of more capacity consumption. Several out-of-band monitoring structures, including simple/non-simple cycles, paths, non-simple trails, and bidirectional trails, etc., have been extensively studied [2, 3, 5, 12, 18, 20, 27, 42, 45, 46, 51].

First, the concept of a simple *monitoring cycle* was proposed in [52]. A simple m-cycle is a supervisory lightpath starting and terminating at the same node, which passes through each on-cycle node exactly once. Later the m-cycle concept was extended to non-simple m-cycles in [45]. In contrast to a simple m-cycle, a non-simple m-cycle is allowed to pass through a node multiple times.

Afterwards, the concept of *monitoring trails (m-trail)* was proposed [40, 44]. It differs from simple and non-simple m-cycles by removing the cycle constraint, and thus an m-trail can be taken as a connected supervisory lightpath with an associated monitor equipped at the destination node of the m-trail. Physical length limits on m-trails were also considered in [28].

Finally, the least constrained monitoring structure the *bidirectional m-trail (bm-trail)* was introduced in [18], where the only constraint on the set of links traversed by the lightpath is that they should be interconnected. In this case, additionally we assume bidirectional optical links in the network, and thus they can be traversed by a directed route using the depth first search (DFS) order. Note that when implementing this route as a single lightpath the nodes may require loop-back switching the optical signals coming from the transmission fiber into its reception fiber.

3.1.1.3 Failure Localization Time

The failure localization speed depends on two factors: the signaling overhead and the failure detection time for each m-trail. The latter mainly depends on the physical length of the lightpath. As for the signaling overhead, there are three frameworks. In the first, the node where the lightpath terminates generates alarms upon any irregularity. The generated alarms are collected and used for failure localization at the central controller. Such alarm dissemination is at the expense of signaling overhead in the control plane, but also makes the failure localization mechanism dependent on efforts not in the optical domain. In such a framework the goal is to achieve *UFL*, which means any SRLG failure can be precisely and instantly identified via monitoring a set of supervisory lightpaths.

We define a lightpath to be *local* to a node if it terminates in the node and, thus its status can be monitored by the node. The failure localization speed can be increased if the monitoring lightpath status information is exchanged among the fewest possible nodes. Ideally there is a single node in which every monitoring lightpath terminates. Such a node is called a *monitoring location* and is said to be capable of achieving *Local Unambiguous Failure Localization (L-UFL)*. In L-UFL after the failure is localized, the monitoring location node should broadcast this information in the network, and initiate the restoration process.

Motivated by the fact that failure localization should be carried out completely in the optical domain without taking any control plane signaling effort, a new framework was proposed recently [38, 39, 47]. It allows each node to inspect the on-off status of the monitoring lightpaths that traverse through the node, which can be done via optical signal tapping. Thus, all the nodes traversed by the monitoring

3.1 Introduction 39

lightpath can share the on-off status of the lightpath. Note that a node can only monitor the links/components of a local lightpath which are upstream to the node. Here the goal is to achieve *NL-UFL—Network-Wide Local UFL* in the network. This is achieved if every node is L-UFL capable.

Using the fact that a node doesn't need to respond to the failure of a distant link in the network, the concept of the NL-UFL approach can be generalized and the failure restoration process can be made faster. This approach is called *Global Neighborhood Failure Localization (G-NFL)*, where each node only localizes a link failure for which it has to respond in the restoration process of an arbitrary shared protection approach, e.g., Failure Dependent Protection in Sect. 2.3.5.

3.1.2 Problem Input

In a general out-of-band failure localization problem the inputs are the following.

1. The *network topology*, which is represented by an undirected graph $G = (V, E)$. The network is assumed to be 2-connected.
2. The *set of SRLGs*, which is denoted by $\mathscr{Z}$. Each SRLG $z \in \mathscr{Z}$ contains one, or multiple links.
3. The required shape of monitoring lightpaths (e.g., m-cycle, m-path, m-trail, and bm-trail).

3.1.2.1 Shape Constraints

Each link e must be assigned with a binary alarm code $a_e = [a_{e,[1]}, a_{e,[2]}, \ldots, a_{e,[b]}]$, where b is the length of the alarm code. The lth binary digit, denoted by $a_{e,[l]}$, is 1 if the lth monitoring lightpath, denoted by T_l, traverses through link e, and 0 otherwise. Note that T_l has to traverse through all the links e with $a_{e,[l]} = 1$ while avoiding to take any link with $a_{e,[l]} = 0$. Depending on the *constraints on the shape* of the monitoring lightpath we have the following conditions on T_l:

bm-trail: T_l must be connected;
m-trail: T_l has an Euler trail, which is a connected subgraph where every node must have an even nodal degree except two: the source and destination node;
m-path: T_l must be an m-trail which passes through each node only at most once;
m-cycle: T_l must be an m-trail and the source and destination node must be the same node.

Further problem specific constraints are introduced in the next sections.

Table 3.1 Classification of hard failure localization techniques

		Hard failures		
		Single-link	Sparse-SRLG	Multiple failures
UFL	m-cycle	Sect. 3.2.1.2		
	m-trail	Sects. 3.2.2 and 3.2.7	Sects. 3.3.5 and 3.3.6	
	bm-trail	Sects. 3.2.3 and 3.2.5	Sect. 3.3.3	Sect. 3.3.1
L-UFL	m-cycle/path	Sect. 4.2.3	[3]	
NL-UFL	bm-trail	Sect. 4.2.4		
	in-band	[23, 33]	[26, 27]	
G-NFL	m-cycle/path	Sect. 6.2		

3.1.2.2 General Objective Function

The cost function in out-of-band monitoring is typically composed of two ingredients:

Monitoring cost, denoted by b, which is the number of monitoring lightpaths, reflects the fault management complexity. A smaller number of monitoring lightpaths results in shorter alarm codes, which further affects the number of alarms flooded in the network when a failure event occurs. In addition to larger fault management cost, a longer alarm code may cause a longer failure recovery time since a network entity has to collect all the necessary alarm signals for making a correct failure localization decision.

Bandwidth cost, denoted by $\|\mathcal{T}\|$, which reflects the additional bandwidth consumption for monitoring. It is also called the *cover length* which is the sum of the lengths of each monitoring lightpath in the solution. The *length* of a monitoring lightpath is often taken as the number of links traversed by the lightpath. In this case the bandwidth cost is nothing else but the total number of one bits in the alarm codes assigned to the links.

Table 3.1 indicates the sections dealing with each sub-problem. Notations are summarized in Table 3.2.

3.2 UFL for Single Failures

3.2.1 Problem Definition

The *UFL constraint* requires every link alarm code to be unique and we also have the shape constraints described in Sect. 3.1.2.1. In most of the previous works [40, 44, 45], the objective was to minimize the weighted sum of the *monitoring cost* and *bandwidth cost*, formally

3.2 UFL for Single Failures

Table 3.2 Notation list

Notation	Description
$G = (V, E)$	Undirected graph representation of the topology
$|V|$	The number of nodes in G
$|E|$	The number of links in G
b	The number of m-trails
$\mathcal{T} = \{T_1, \ldots, T_b\}$	A solution with b (b)m-trails
T_i	The ith (b)m-trail, which is a set of link in G
$|T_i| = t_i$	The length in hops of ith m-trail
$\|\mathcal{T}\| = \sum_{i=1}^{b} |T_i|$	The total cover length
A	The alarm code table (ACT)
a_e	The alarm code of link $e \in E$
$a_{e,[j]}$	The jth bit of the alarm code of link $e \in E$
$a_{\{e_1,e_2\}}$	The alarm code of SRLG $\{e_1, e_2\}$
$a_{\{e_1,e_2\},[j]}$	The jth bit of the alarm code of SRLG $\{e_1, e_2\}$
L_j	The set of links with "1" in the jth bit position

$$\mathcal{C} = \gamma \times (\# \ of \ m\text{-}trails) + cover \ length = \gamma b + \|\mathcal{T}\|, \tag{3.1}$$

where $\mathcal{C}$ denotes the total cost of the solution.

In more theoretical studies on bm-trails the bandwidth cost is usually ignored, and only the number of bm-trails is considered. Some studies take out-of-band monitoring problems as a network dimensioning problem, and take the capital expenditures instead of operating expenses [12]. In those studies the main cost is the number of transmitters.

3.2.1.1 A UFL Example

Figure 3.1 shows an example of a UFL m-trail solution to the network for localizing any single-link failure, and its *alarm code table* (ACT) is also shown. The ACT stores the alarm code of each link (e.g., link (v_8, v_9) is assigned the alarm code [10001]), which further defines how the five m-trails (i.e., $T_1, \ldots, T_5$) should be routed. Each row of the ACT should be unique to achieve UFL. Here, T_j has to traverse through all the links with the jth bit of the alarm code "1" while avoiding to take any link with the jth bit of its alarm code "0." By reading the status of the five m-trails, any link failure can be unambiguously localized. For example, the darkness of T_1 and T_5 and no other T_i implies the failure of link (v_8, v_9).

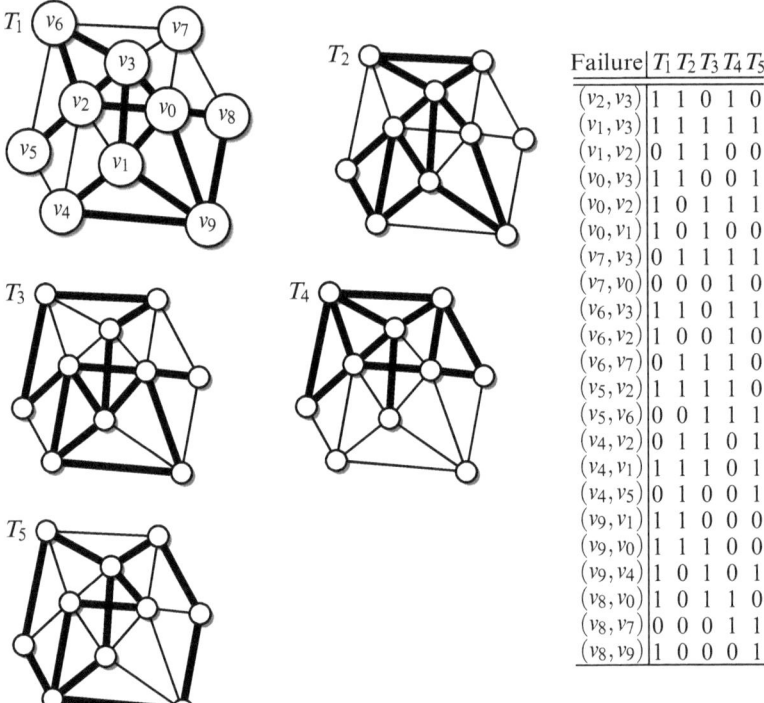

Fig. 3.1 Unambiguous failure localization (UFL) based on m-trails. The solution has $b = 5$ and $\|\mathcal{T}\| = 64$

3.2.1.2 UFL for Single Failure with M-Cycles

To distinguish the failure of two links adjacent with a degree-2 node v, we need a monitoring lightpath that terminates in v, which is clearly not possible with m-cycles. Since network topologies often have degree-2 nodes, most of the recent papers deal with more relaxed shape constraints. The ILP for optimal design was formulated in [45, 52].

3.2.2 Lower and Upper Bounds on the Number of (B)M-Trails

The theoretical lower bound on the number of m-trails is

$$b \geq \lceil \log_2 (|E| + 1) \rceil,$$

3.2 UFL for Single Failures

Table 3.3 The best known lower and upper bounds on the number of (b)m-trails of different graphs

Topology	Shape	Lower bound	Upper bound	
			In [18]	In the book
Ring	bm-trail	$\lceil |E|/2 \rceil$	$\lceil |E|/2 \rceil$	Theorem 3.1
Graph without degree-2 nodes	bm-trail	$\frac{|E|}{12}$, Theorem 3.2		
2D grid	bm-trail	$\lceil \log_2(|E|+1) \rceil$	$\approx 3\lceil \log_2 |E| \rceil$	Tight + 5, Theorem 3.6
Well connected	bm-trail	$\lceil \log_2(|E|+1) \rceil$	$\approx 2\lceil \log_2 |E| \rceil - 1$	Tight, Theorem 3.4
$C_{1,2}$ circulant	bm-trail	$\lceil \log_2(|E|+1) \rceil$		Tight, Theorem 3.7
Fully connected	m-trail	$\lceil \log_2(|E|+1) \rceil$	$\approx 2\lceil \log_2 |E| \rceil - 1$	Tight + 4, Theorem 3.3
Chocolate bar	m-trail	$\lceil \log_2(|E|+1) \rceil$		Tight + 0.42, Theorem 3.5

since there are $|E|$ single failure states plus the no failure state, and the 2^b potential alarm codes must distinguish these, giving that $2^b \geq |E| + 1$.

Table 3.3 summarizes the best known lower and upper bounds on the number of (b)m-trails reported in the literature for several special graphs. For ring topologies, the number of optimal bm-trails is exactly $\lceil |E|/2 \rceil$, which was proved first in [18], and later for m-trails in [37]. Note that in ring topologies each bm-trail should only be a simple path.

The study [18] developed a construction for any graph which contains two edge-disjoint spanning trees, where an upper bound of $2 \cdot \lceil \log_2 |E| \rceil - 1$ bm-trails can be achieved. The key idea of the construction is to categorize the links in the topology into two disjoint sets E_1 and E_2 of similar sizes, where $E_1 \cup E_2 = E$, $E_1 \cap E_2 = \emptyset$, and each set contains a spanning tree. We shall generate alarm codes of length $\lceil \log_2 |E_1| \rceil + \lceil \log_2 |E_2| \rceil$ for the links in E. The links in E_1 will have unique codes in the first $\lceil \log_2 |E_1| \rceil$ bits, and similarly the links in E_2 are coded uniquely in the last $\lceil \log_2 |E_2| \rceil$ bits. At this point every link has a unique alarm code, irrespective of the values in the bits in the alarm code that has not yet been specified. These unspecified bits can be used to make the resulting test sets connected (actually spanning the whole V) and form bm-trails. Finally, we add one additional bm-trail covering every link in E_1 and none in E_2, which can identify if the failed link belongs to E_1 or E_2. In such a way, each link has a unique alarm code with a length:

$$\lceil \log_2 |E_1| \rceil + \lceil \log_2 |E_2| \rceil + 1 \approx \lceil \log_2 \frac{|E|}{2} \rceil + \lceil \log_2 \frac{|E|}{2} \rceil + 1 = 2 \cdot \lceil \log_2 |E| \rceil - 1. \tag{3.2}$$

We refine this idea further in Sect. 3.2.3.

Nash-Williams and Tutte [11] showed that every $2k$-connected graph[1] has k link-disjoint spanning trees. Note that the disjoint spanning trees can be found in $O(|V||E|\log\frac{|E|}{|V|})$ time [15]. As a result, every 4-connected graph has 2 link-disjoint spanning trees, thus the proof is valid for complete graphs with more than 3 nodes.[2] For two-dimensional square grid lattices, on the other hand, a similar technique was developed in [18] which results in $2 + 6 \cdot \lceil \log_2(n+1) \rceil$ as an upper bound on the number of bm-trails, where the graph has $(n+1) \times (n+1)$ nodes. In fact in the square grid lattice, due to $|E| = 2n^2 + 2n < 2(n+1)^2$ we have

$$2 + 6 \cdot \lceil \log_2(n+1) \rceil \leq 2 + 6 \lceil \log_2 \sqrt{\frac{|E|}{2}} \rceil \approx 3 \lceil \log_2 |E| \rceil, \quad (3.3)$$

which is about three times of the theoretical lower bound: $\lceil \log_2(|E|+1) \rceil$.

In Sect. 3.4.1 an observation made from extensive simulations on thousands of general topologies is that, the m-trail solution on a topology without degree-2 nodes can achieve the theoretical lower bound of $1 + \lceil \log_2(|E|+1) \rceil$ provided sufficient running time for the construction. This was disproved by an example in Sect. 3.2.2.2. In this section we show a suite of polynomial-time deterministic constructions toward optimal (or essentially optimal) solutions for the (b)m-trail UFL problem.

3.2.2.1 Ring Networks

The lower bound on the number of m-trails in ring network is proved as follows [18].

A ring is a network on vertices $v_1, \ldots v_n$ whose edges (links) are (v_1, v_2), $(v_2, v_3), \ldots, (v_{n-1}, v_n), (v_n, v_1)$. Here n is the length of the ring.

Theorem 3.1. *A ring topology of more than 4 nodes needs $\lceil |E|/2 \rceil$ m-trails for single-link UFL.*

Proof. We divide the proof into two claims: (1) a ring topology needs at least $\lceil |E|/2 \rceil$ m-trails for single-link UFL, and (2) a ring topology needs no more than $\lceil |E|/2 \rceil$ m-trails for single-link UFL.

[Proof of claim (1)] Let e and f be two links with a common adjacent node v, as shown in Fig. 3.2. In order to unambiguously identify failure between these two links, there must be an m-trail that passes through link e but not link f (or vice versa). Since v has degree two, this can only happen if an m-trail terminates at node v. It is clear that in a ring topology, a number of $|E|$ adjacent link-pairs can be found, and each m-trail has two terminating nodes. Therefore, it requires at least $\lceil |E|/2 \rceil$ m-trails to achieve that all the n nodes are endpoints of an m-trail.

---

[1]There does not exist a set of at most $2k-1$ edges whose removal disconnects the graph.
[2]Based on a similar approach, an upper bound $(6 + \lceil \log_2(|E|+1) \rceil)$ for the m-trail formation problem is proved in Sect. 3.2.2.3.

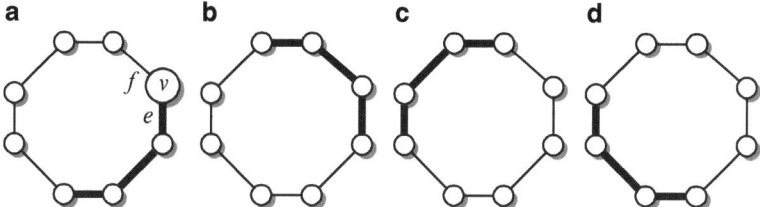

Fig. 3.2 Optimal M-trail assignment of an 8-node ring. (**a**) M-trail T_1, (**b**) M-trail T_2, (**c**) M-trail T_3, (**d**) M-trail T_4

[Proof of claim (2)] In a ring topology, every single-link failure can be unambiguously identified in such a way that each m-trail is 3-hop in length and overlaps with its two neighbor m-trails by one hop, as shown in Fig. 3.2. If the ring has an odd number of nodes, the last m-trail must be a 2-hop m-trail. Thus, the network needs up to $\lceil |E|/2 \rceil$ m-trails for achieving single-link UFL.

3.2.2.2 Lower Bound on the Number of M-Trails in General Graphs Without any Degree-2 Nodes

Theorem 3.1 can be extended to the scenario of general Euler graphs in the derivation of an upper bound on the number of m-trails. Next let us give a lower bound on the number of m-trails in some "bad" two-edge-connected graphs. Clearly we have the lower bound of $\lceil \log_2(|E|+1) \rceil$ due to the binary coding mechanism, which accounts for the fact that it takes $\lceil \log_2(|E|+1) \rceil$ bits to unambiguously identify $|E|$ different states (if "000...0" is not considered). In the following paragraphs we will demonstrate another lower bound on the number of m-trails of two-edge-connected topologies that works in parallel with the lower bound by $\log_2(|E|+1)$.

Assume that we have a set of node-disjoint graphs $G_1, G_2, \ldots, G_n$. Let the node set of G be the union of the node sets of G_i for $i = 1, \ldots n$. The edges of G are the edges of G_i and the connecting links $e_1, \ldots e_n$, where e_i connects a node of G_i to a node of G_{i+1} for $i = 1, \ldots n-1$, and e_n connects a node of G_n to a node of G_1. Clearly G is a two-edge-connected graph if each G_i is two-edge-connected. The set of edges $\bar{E} = \{e_1, \ldots e_n\}$ is called the *separating set*. The edges from $\bar{E}$ are called *separating links*. In the example of Fig. 3.3b we may assume that $n = 4$, and the gray links are the separating links. We shall consider m-trails in G. We call a component G_i a *boundary* of a trail T, if T includes exactly one of the separating edges incident to G_i.

Theorem 3.2. *At least $\lceil \frac{|\bar{E}|}{2} \rceil = \lceil \frac{n}{2} \rceil$ m-trails are needed to establish single-link UFL in the graph G above.*

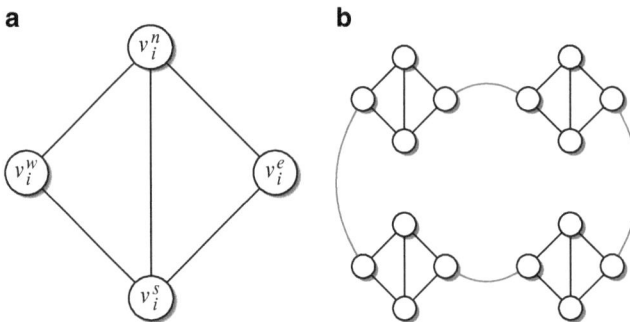

Fig. 3.3 The structure of each component. (**a**) The ith component, (**b**) a counter example with $|\bar{E}| = 4$

Proof. First we show that any m-trail T has at most two boundary components. Indeed, contract every component G_i into a single node. This transforms G into a ring, while the image of T will still be a connected subgraph. Connected subgraphs in a ring have at most two points of degree one. This implies that T has at most two boundary components.

Second, we establish that every component G_i must serve as a boundary for some m-trail T. Indeed, let e and f be the separating edges incident to G_i. As the collection of our m-trails provides UFL for single-link failures, there must be a trail T which contains f and does not contain e, or conversely, one which contains e but not f. In both cases G_i must be a boundary for T.

With the above two claims, we know that each m-trail has at most two boundary components, and each component must be the boundary of at least a single m-trail. Therefore, the number of m-trails in the topology is at least $\frac{|\bar{E}|}{2}$.

With Theorem 3.2, the logarithmic relation between the number of m-trails and network size could be broken due to the presence of $\bar{E}$. Therefore, we can easily see that *an m-trail solution for a two-edge-connected topology with $c + \log_2(|E|)$ m-trails may not exist even if the topology does not contain any degree-2 nodes*, where c is a small positive constant. Let us define a network topology as *logarithmically proper* if an m-trail solution for the single-link failure localization problem can be found with $c + \log_2(|E|)$ m-trails. Obviously, a fully meshed topology and grid topology are *logarithmically proper*, which can be covered with $c + \log_2(|E|)$ m-trails (according to the construction in Sects. 3.2.2.3 and 3.2.4, respectively), while a ring topology is not (according to Theorem 3.1). The topology in Fig. 3.3b has $|\bar{E}| = 4$ components although without any degree-2 node, and the structure of the components is illustrated in Fig. 3.3a. The number of m-trails for the graphs like on Fig. 3.3b has the following lower bound:

$$b \geq \frac{|\bar{E}|}{2} = \frac{|E|}{12} \tag{3.4}$$

3.2 UFL for Single Failures

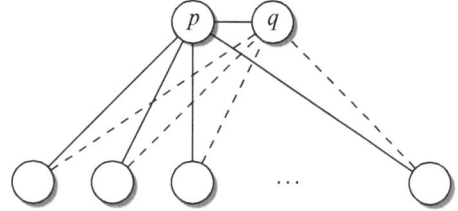

Fig. 3.4 Subgraph G_p is drawn with *solid lines* and G_q with *broken lines*, while G' contains all the rest of the links of the complete graph

Equation (3.4) holds because each component (as shown in Fig. 3.3b) along with a separating link totally has 6 links, which yields $|E| = 6 \cdot |\bar{E}|$.

3.2.2.3 Near Optimal Construction for Fully Meshed Networks

The subsection introduces a deterministic polynomial time construction of an m-trail solution for fully meshed topologies (i.e., complete graphs) that employs $4 + \lceil \log_2 (|E| + 1) \rceil$ m-trails for UFL. Theorem 3.3 formulates the correctness of the proposed construction. Among the six steps in the construction, Step (1) is for initialization, Step (2)–Step (5) are to ensure the code uniqueness of each link, and Step (6) is for m-trail formation.

Input: The complete graph $G = (E, V)$
Result: UFL solution with $4 + \lceil \log_2 (|E| + 1) \rceil$ m-trails for $|V| \geq 7$

Step (1) Let $\mathfrak{B} = \lceil \log_2(|E| + 1) \rceil$ be the theoretical lower bound on the number of m-trails. $G = (E, V)$ is first decomposed into three link-disjoint subgraphs denoted as $G' = (E', V)$, $G_p = (E_p, V)$, and $G_q = (E_q, V)$, such that p and q are two different nodes of V, while E_p consists of every link adjacent to node p; and similarly E_q consists of every link adjacent to node q except the link (p, q). All the other links and nodes $v \in V \setminus \{p, q\}$ in G' form a complete graph with $|V| - 2$ nodes. Thus, we have $|E_p| = |V| - 1$, $|E_q| = |V| - 2$, and $|E'| = |E| - 2|V| + 3$. As shown in Fig. 3.4, G_p and G_q both have the shape of a star with central nodes p and q, where $p \neq q$.

Step (2) We first allocate two m-trails, denoted as $T_{\mathfrak{B}+3}$ and $T_{\mathfrak{B}+4}$, to distinguish whether a link of G belongs to G', G_p, or G_q. As shown in Fig. 3.5, one example to achieve the above is to route the m-trail $T_{\mathfrak{B}+3}$ through all the links in $G_p \cup G_q$ while $T_{\mathfrak{B}+4}$ over all the links of G_q (and some links of G'). $T_{\mathfrak{B}+3}$ is a valid m-trail (which admits an Euler trail from p to q) because the nodal degree of each node along $T_{\mathfrak{B}+3}$ is always even except possibly at p and q. Since G_q is a star topology, the routing of $T_{\mathfrak{B}+4}$ needs some links from G' until the Euler property is met. An example of such a link set is the edge set in G' of a maximum matching.

Fig. 3.5 An example $T_{\mathcal{B}+3}$ and $T_{\mathcal{B}+4}$, where $T_{\mathcal{B}+3}$ is drawn with *solid lines* and $T_{\mathcal{B}+4}$ with *broken lines*

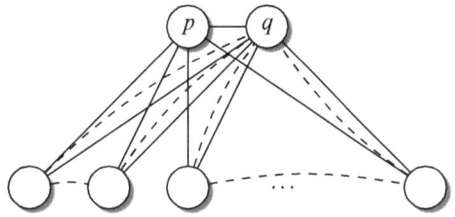

Note that $T_{\mathcal{B}+4}$ is used to distinguish the links in G_p from those in G_q, and $T_{\mathcal{B}+3}$ is to distinguish links in G' from those in G_p or in G_q. Therefore with $T_{\mathcal{B}+3}$ and $T_{\mathcal{B}+4}$, the overall UFL can be achieved provided that UFL can be achieved separately in each of the three subgraphs G', G_p, G_q. This will be done in the following steps.

Step (3) Unique nonzero binary codes of length $\lceil \frac{\mathcal{B}+1}{2} \rceil$ bits are generated for the links in E_p. This can be done because

$$2^{\lceil \frac{\mathcal{B}+1}{2} \rceil} \geq 2^{\frac{\mathcal{B}+1}{2}} = \sqrt{2 \cdot 2^{\mathcal{B}}} \geq \sqrt{2|E|} = \sqrt{2 \frac{|V|(|V|-1)}{2}} > \sqrt{(|V|-1)^2} = |V|-1. \tag{3.5}$$

The codes generated here are called *core codes* for E_p, and each of the codes serves as a $\lceil \frac{b+1}{2} \rceil$ bit-long prefix for the alarm code assigned to a link of E_p. The structure of the codes can be expressed as:

m-trails: for links in E_p	$T_1 \ldots T_{\lceil \frac{\mathcal{B}+1}{2} \rceil}$ core code	$T_{\lceil \frac{\mathcal{B}+1}{2} \rceil + 1} \ldots T_{\mathcal{B}+2}$ $x \ldots xx$	$T_{\mathcal{B}+3}$ 1	$T_{\mathcal{B}+4}$ 0

where x denotes the yet undefined bits. To the link (p,q) we associate the code $[1, \ldots, 1, 0]$ which has $\mathcal{B}+3$ "1" bits, and one "0" bit.

Step (4) Next unique nonzero binary codes with $\lceil \frac{\mathcal{B}+1}{2} \rceil$ bits are generated for the links in E_q. The codes generated are called *core codes* for E_q, and each of the codes serves as a $\lceil \frac{\mathcal{B}+1}{2} \rceil$ bit-long substring for the alarm code assigned to each link in E_q. The structure of the codes can be expressed as:

m-trails: for links in E_q	$T_1 \ldots T_{\lceil \frac{\mathcal{B}+1}{2} \rceil}$ $x \ldots xx$	$T_{\lceil \frac{\mathcal{B}+3}{2} \rceil} \ldots T_{\mathcal{B}+2}$ core code	$T_{\mathcal{B}+3}$ 1	$T_{\mathcal{B}+4}$ 1

Step (5) Unique nonzero codes with $\mathcal{B}+2$ bits are generated as the *core codes* for the links in E'. Note that this can easily be done since $|E| < 2^{\mathcal{B}}$. The generated unique codes are assigned to the links in a structured manner described as follows. Recall that E' is a complete graph on $|V|-2 \geq 5$ nodes. We identify two link-disjoint Hamiltonian cycles on the links of E' (e.g., by way

3.2 UFL for Single Failures

of Walecki's construction [4, 22]), denoted by H_1 and H_2, which cover every node except $\{p, q\}$. For each link in H_1, "1" is assigned to each bit at the bit positions $1, \ldots, \lceil \frac{\mathfrak{B}+1}{2} \rceil$. Note that according to Eq. (3.5), at least $|V| - 1$ such codes exist. Similarly, for each link of H_2, "1" is assigned to each bit at the bit positions $\lceil \frac{\mathfrak{B}+3}{2} \rceil, \ldots, \mathfrak{B} + 2$. The format of the codes for the links of E' is given as follows.

m-trails:	$T_1 \ldots T_{\lceil \frac{\mathfrak{B}+1}{2} \rceil}$	$T_{\lceil \frac{\mathfrak{B}+3}{2} \rceil} \ldots T_{\mathfrak{B}+1}$	$T_{\mathfrak{B}+2}$	$T_{\mathfrak{B}+3}$	$T_{\mathfrak{B}+4}$
for links in H_1	$11 \ldots 1$	code fragment		0	x
H_2	code frag.	$11 \ldots 1$	1	0	x
$E' \setminus H_1 \cup H_2$	core code in $\mathfrak{B} + 1$ bits		x	0	x

Step (6) After Step (2)–Step (5), we can identify the link sets L_j, $1 \leq j \leq \mathfrak{B} + 2$, which contains the links with "1" in the jth bit position in G. Let the link set contain the links with an undefined bit at the jth bit position be denoted as L_j^x. Now our objective in this step is to extend L_j by using some of those links in L_j^x such that a valid m-trail T_j can be formed. This equivalently determines the bits of x in each link.

To ensure that L_j forms an Eulerian trail (either open or closed), we sequentially check the vertices $v \in V \setminus \{p, q\}$ to see if the degree of each v is odd or even in the current L_j. If v has an odd nodal degree, (v, q) is added to L_j if $j \leq \lceil \frac{\mathfrak{B}+1}{2} \rceil$, and (v, p) is added to L_j if $\lceil \frac{\mathfrak{B}+1}{2} \rceil < j \leq \mathfrak{B} + 2$. Therefore, we can make sure that only p and q may have an odd degree in L_j.

Then we check L_j to see if it spans a connected graph. If not, then due to the presence of one of the cycles H_1 or H_2, (p, q) is in L_j and it must be an isolated edge. In this case we simply add a link (v, p) into L_j for $v \in V \setminus \{p, q\}$ (or (v, q), respectively). The resulting graph must have an Euler trail because the odd-degree nodes must be in the set $\{v, p, q\}$.

Theorem 3.3. *The proposed construction on a complete graph gives at most*

$$b = 4 + \lceil \log_2 (|E| + 1) \rceil$$

m-trails to achieve UFL for $|V| \geq 7$.

Proof. The proof of the construction is divided into two parts: (a) the code uniqueness of each link, and (b) the successful formation of an m-trail for each bit position. As for the latter, we will show that all the links with "1" in the jth bit position are connected to form a valid m-trail.

For part (a), the links in each subgraph have unique codes due to the intrinsic nature of the core code generation in each subgraph, which were presented in Step (3)–Step (5). Also by Step (2), the $(\mathfrak{B} + 3)$th and $(\mathfrak{B} + 4)$th bit positions are used to distinguish the links of the three subgraphs G', G_p, G_q. Therefore, the code uniqueness of each link can be ensured. For part (b), Step (6) ensures that each link set L_j are all connected with no more than 2 nodes with an odd nodal degree. Note that

$$b = \mathcal{B} + 2 - \lceil \frac{\mathcal{B}+1}{2} \rceil = \mathcal{B} + 1 - \lceil \frac{\mathcal{B}+1}{2} \rceil + 1 = \lfloor \frac{\mathcal{B}+1}{2} \rfloor + 1 \geq \lceil \frac{\mathcal{B}+1}{2} \rceil,$$

hence for $1 \leq j \leq \lceil \frac{\mathcal{B}+1}{2} \rceil$ the edges of G_q, while for $\lceil \frac{\mathcal{B}+3}{2} \rceil \leq j \leq \mathcal{B} + 2$ the edges of G_p can be used. Also note that L_j spans a connected graph on $V \setminus \{p, q\}$, due to the presence of the Hamiltonian cycles H_1 and H_2 as described in Step (5). Therefore, each L_j, $1 \leq j \leq (\mathcal{B} + 4)$, will form a valid m-trail.

With all the above, we proved that the proposed construction has each link coded with $b = (\mathcal{B} + 4)$ bits. This gives $(\mathcal{B} + 4)$ valid m-trails for achieving UFL in the fully meshed (or complete) graph G.

Note that the proposed m-trail solution for fully meshed topologies solves a special case of the problem addressed in [18] by Algorithm 1, and thus it improves the $O(\log_2 |E|)$ construction (Theorem 2 of [18]) to $O(1) + \log_2 |E|$.

3.2.3 An Optimal BM-Trail Solution in Densely Meshed Graphs

We shall need a simple inequality.

Lemma 3.1. *The following inequality holds for every positive integer $b \geq 3$:*

$$2 \cdot \lfloor \frac{2^b - 1}{b} \rfloor \geq \lfloor \frac{2^{b+1} - 1}{b+1} \rfloor \geq \lceil \frac{2^b - 1}{b} \rceil. \tag{3.6}$$

Proof. For the first inequality one can readily check that it holds for $b = 3, 4, 5$. Note also that the inequality fails for $b = 2$. We have

$$2 \cdot \lfloor \frac{2^b - 1}{b} \rfloor - \frac{2^{b+1} - 1}{b+1} \geq 2 \cdot \left(\frac{2^b - 1}{b} - 1 \right) - \frac{2^{b+1} - 1}{b+1}. \tag{3.7}$$

After clearing denominators, the nonnegativity of the above quantity for $b \geq 6$ is equivalent to $2^{b+1} - (2b^2 + 3b + 2) \geq 0$. But for $b \geq 4$ we have $3b + 2 < 3b + b = 4b \leq b^2$, hence it suffices to see that $2^{b+1} - 3b^2 \geq 0$, or $f(b) := \frac{2^{b+1}}{3b^2} \geq 1$. We have $f(6) = \frac{128}{108} > 1$. Moreover, for every real $b \geq 3$,

$$\frac{f(b+1)}{f(b)} = 2 \left(1 - \frac{1}{b+1} \right)^2 \geq 2 \left(1 - \frac{1}{4} \right)^2 = \frac{18}{16} > 1,$$

3.2 UFL for Single Failures

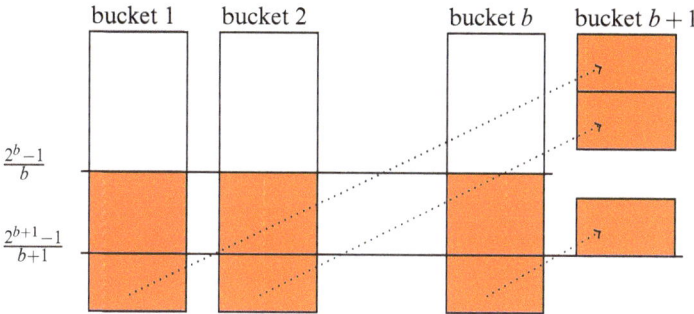

Fig. 3.6 Example of the construction in the proof of Lemma 3.2

It implies that $f(b) > 1$ whenever $b \geq 6$ is an integer. The second inequality holds because $\frac{2^b-1}{b}$ is a convex increasing function, and at $b = 4$ the difference is $\frac{2^5-1}{5} - \frac{2^4-1}{4} = 2.45 > 1$.

Next let us prove a lemma which is an important building block for the subsequent description on the proposed construction and its proof.

Lemma 3.2. *The nonzero binary codewords of length b can be distributed into b buckets, where the ith bucket contains codewords only with 1 for the ith bit, and the size of each bucket is at least $\lfloor \frac{2^b-1}{b} \rfloor$ and at most $\lceil \frac{2^b-1}{b} \rceil$.*

Proof. The proof is inductive, and we will give a recursive construction for such a distribution of codewords. See Fig. 3.6 as an illustration of each recursive step.

Clearly, for $b = 1, 2$ the statement trivially holds. Let us assume that the codewords of length b are already distributed into b buckets, where the ith bucket has words only with 1 for the ith bit, and the size of each bucket is at least $\lfloor \frac{2^b-1}{b} \rfloor$ and at most $\lceil \frac{2^b-1}{b} \rceil$. We define such a distribution as an *almost uniform distribution of b bits*.

Next, we consider the nonzero codewords of length $b + 1$, and prove that there exists an almost uniform distribution of $b + 1$ bits. Clearly we can distribute the $2^b - 1$ codewords with 0 in the $(b + 1)$th bit such that the first b bits are distributed almost uniformly (inductive hypothesis): the first b buckets are filled up with at least $\lfloor \frac{2^b-1}{b} \rfloor$ and at most $\lceil \frac{2^b-1}{b} \rceil$ codewords. At the end these buckets must have at least $\lfloor \frac{2^{b+1}-1}{b+1} \rfloor$ and at most $\lceil \frac{2^{b+1}-1}{b+1} \rceil$ codewords.

Next, let us consider the rest of the codewords. Obviously, any of them can be placed into the $(b + 1)$th bucket, because they all have 1 bit at position $b + 1$. The codeword which has 1 at the $(b + 1)$th position and 0 in the rest positions (i.e., $100\ldots0$) must be placed into the $(b+1)$th bucket. The remaining $2^b - 1$ codewords can be distributed by the first b bits almost uniformly into the b buckets. In such a way, each bucket has at least $2 \cdot \lfloor \frac{2^b-1}{b} \rfloor$ codewords, which is at least $\lfloor \frac{2^{b+1}-1}{b+1} \rfloor$ according to Lemma 3.1. Some of the newly added codewords must be moved to the $(b+1)$th bucket, until every bucket has at least $\lfloor \frac{2^{b+1}-1}{b+1} \rfloor$ and at most $\lceil \frac{2^{b+1}-1}{b+1} \rceil$. Such an action is always possible. This is argued as follows: first, codewords are moved from each of the first b buckets to the $(b+1)$th bucket so that every bucket among the first b has at most $\lceil \frac{2^{b+1}-1}{b+1} \rceil$ elements and at least $\lfloor \frac{2^{b+1}-1}{b+1} \rfloor$ elements. In case the $(b+1)$th bucket has less than $\lfloor \frac{2^{b+1}-1}{b+1} \rfloor$ codewords, one more codeword from each bucket is further moved to the $(b+1)$th bucket until it has $\lfloor \frac{2^{b+1}-1}{b+1} \rfloor$ codewords. Such a process will not get stuck at a position in which one bucket has less than $\lfloor \frac{2^{b+1}-1}{b+1} \rfloor$ codewords while all the others have this number, because the total number of nonzero codewords is $2^{b+1}-1$. In such a way, every bucket has at least $\lfloor \frac{2^{b+1}-1}{b+1} \rfloor$ and at most $\lceil \frac{2^{b+1}-1}{b+1} \rceil$ codewords. Thus, we proved Lemma 3.2.

Theorem 3.4. *Let $G = (V, E)$ be a $2 \cdot \lceil \log_2(|E|+1) \rceil$ connected graph. Then $G = (E, V)$ can be optimally covered with $\lceil \log_2(|E|+1) \rceil$ bm-trails to achieve single-link UFL.*

Proof. Let $b = \lceil \log_2(|E|+1) \rceil$. Clearly at least b bm-trails are required for UFL in a graph with $|E|$ links. In the following we will show that b is also the upper bound. Our goal for the proof of the theorem is to come up with a construction that achieves the theoretical lower bound, and then we will prove the correctness of the construction.

3.2.3.1 The Proposed Construction

Recall that the goal of the bm-trail formation process is to assign a binary alarm code to each link so that T_i is a connected subgraph, where T_i denotes the set of links with alarm codes having 1 at ith position for $i = 1, \ldots, b$. This can be ensured if each T_i has a spanning tree as a subgraph. Since every $2k$-edge-connected graph has k edge-disjoint spanning trees[3] [29, 41], the construction can achieve the desired bound if the graph is $2 \cdot b$ connected, which is sufficient to yield $b = \lceil \log_2(|E|+1) \rceil$ edge-disjoint spanning trees. Let S_i denote the ith spanning tree, where $i = 1, \ldots, b$, and the spanning trees are all disjoint (i.e., $S_i \cap S_j \equiv \emptyset$ if $i \neq j$).

According to Lemma 3.2, the $2^b - 1$ nonzero codewords of b bits in length can be grouped into b buckets of size at least $\lfloor \frac{2^b-1}{b} \rfloor$, where the ith bucket has alarm codes where the ith bit is "1." Our construction simply assigns the codes of the ith bucket to the edges of the ith spanning tree S_i, while the remaining edges which are not in the 1st, $\ldots$, bth spanning trees, namely $E \setminus \{S_1 \cup S_2 \cup \cdots \cup S_b\}$, will be assigned to the left and unused codes arbitrarily. This finishes the construction.

---

[3]Note that such disjoint spanning trees can be found in $O(|V||E|\log \frac{|E|}{|V|})$ time [15].

3.2 UFL for Single Failures

3.2.3.2 Correctness of the Constructed BM-Trail Solution

Since T_i contains S_i, each bm-trail must be connected and span the whole graph G. Besides, each link has a unique alarm code because nonzero unique codewords were assigned to the links of the graph. To conclude the proof we need to show that each bucket has at least $|V| - 1$ codewords. From the inequalities

$$b \cdot (|V| - 1) \le |E| \le 2^b - 1,$$

we see that each bucket has at least

$$|V| - 1 \le \lfloor \frac{2^b - 1}{b} \rfloor$$

elements. Thus, we proved Theorem 3.4.

The theorem is applicable to complete graphs with 18 nodes because they have $\frac{18 \cdot 17}{2} = 153$ links that can be uniquely coded in 8 bits. In this case the graph is at least 16-connected.

3.2.4 An Optimal M-Trail Solution for Chocolate Bar Graphs

Next we consider general 2D grids denoted by $S_{m,n}$, where m and n correspond to the number of links in the vertical and horizontal direction, respectively. Harvey et al. [18] provided a $3 \lceil \log_2 |E| \rceil$ upper bound on the number of bm-trails according to Eq. (3.3) in the case of $m = n$.

In the next section, we generalize the results of [18] and investigate the scenario of 2D grid graphs with arbitrary m and n. We give a novel polynomial-time deterministic construction that requires no more than $3 + \lceil \log_2 (|E| + 1) \rceil$ bm-trails. We first solve the m-trail allocation problem (MAP) for a special case of $S_{m,n}$ with either $n = 1$ or $m = 1$ (called as *chocolate bar graphs*); and then a solution for bm-trails on general 2D grid topologies is developed based on the chocolate bar solution.

A general chocolate bar graph is denoted as $C_n(V, E)$, which has $|V| = 2n + 2$ vertices denoted as $v_{1,0}, \ldots, v_{1,n}$ (the lower points), and $v_{0,1}, \ldots, v_{0,n}$ (the upper points). Figure 3.7a shows an example of a chocolate bar with $n = 6$. For link set E, we have *lower horizontal links* $(v_{1,i}, v_{1,i+1}) \in E$, *upper horizontal links* $(v_{0,i}, v_{0,i+1}) \in E$ for $i = 0, \ldots, n-1$, and the *middle vertical links* $(v_{0,i}, v_{1,i}) \in E$ whenever $i = 0, \ldots, n$.

Theorem 3.5. *For a chocolate bar graph $C_n(V, E)$ $b = \lceil \log_2(n + 1) \rceil + 2$ m-trails achieve single-link UFL for $b > 2$, which is at most $\lceil 0.42 + \log_2 (|E| + 2) \rceil$ m-trails.*

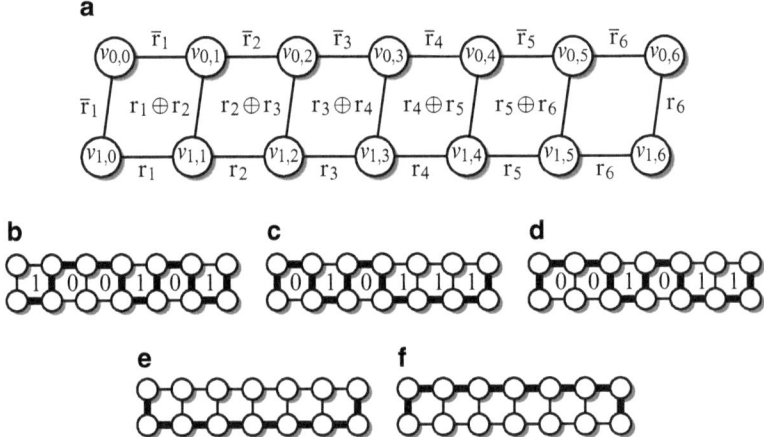

Fig. 3.7 An example of a chocolate bar graph and the corresponding optimal solution for m-trails. The bit of each bit position is drawn in each 1×1 rectangle. The $\mathbf{r}_1, \mathbf{r}_2, \ldots \mathbf{r}_n$ codes assigned to the links are listed in Table 3.4. (**a**) The graph topology, (**b**) the links of T_1, (**c**) the links of T_2, (**d**) the links of T_3, (**e**) the links of $T_{\mathfrak{B}+1}$, (**f**) the links of $T_{\mathfrak{B}+2}$.

Proof. The proof is developed by way of a polynomial-time deterministic construction composed of two steps. We will first introduce the construction, and then explain in detail how the construction can achieve the desired bound on the number of m-trails.

3.2.4.1 Alarm Code Assignment for the Chocolate Bar Graph

Let us assign binary alarm codes to the links of $C_n(V, E)$ in the following way (see also Fig. 3.7). We first generate n bitvectors $\mathbf{r}_1, \mathbf{r}_2, \ldots \mathbf{r}_n$ of length $\mathfrak{B} := \lceil \log_2(n+1) \rceil$, where $\mathbf{r}_{i+1}$ is assigned to a lower horizontal link $(v_{1,i}, v_{1,i+1}) \in E$, where $i = 0, \ldots, n-1$. The generation of these codes is provided in Lemma 3.3. On the other hand, to the higher horizontal link $(v_{0,i}, v_{0,i+1}) \in E$ we assign the bitwise complement of $\mathbf{r}_{i+1}$, denoted by $\bar{\mathbf{r}}^{i+1} = \mathbf{r}_{i+1} \oplus \mathbf{1}$ where $\oplus$ stands for the bitwise modulo 2 addition (XOR) and $\mathbf{1}$ is the all 1 vector of length $\mathfrak{B}$. Also to the middle vertical link $(v_{1,i}, v_{0,i})$ we assign the bitvector $\mathbf{r}_i \oplus \mathbf{r}_{i+1}$ for $i = 1, \ldots, n-1$. Finally to the link $(v_{1,0}, v_{0,0})$ bitvector $\mathbf{r}_1 \oplus \mathbf{1}$ is assigned, and to the link $(v_{1,n}, v_{0,n})$ we attach $\mathbf{r}_n$.

In choosing the list of bitvectors $\mathbf{r}_i$, for $i = 1, \ldots, n$, we make the following three assumptions:

(A1) The vectors $\mathbf{r}_i$ are pairwise different for $i = 1, \ldots, n$.
(A2) The vectors $\mathbf{r}_i \oplus \mathbf{r}_{i+1}$ are all nonzero and pairwise different for $i = 1, \ldots, n-1$.
(A3) The first bits of the vectors $\mathbf{r}_1$ and $\mathbf{r}_n$ are the same.

3.2 UFL for Single Failures

The following statement provides an approach to construct $n \leq 2^{\mathfrak{B}} - 1$ bitvectors $\mathbf{r}_i$ which satisfy the requirements (A1), (A2), (A3).

A Brief Introduction to Galois Fields

In the arithmetic of ordinary numbers there are infinitely many numbers, while the fields $\mathbb{F}_{2^b}$ have only 2^b elements. However, the operations of addition, subtraction, multiplication, and division (except division by zero) may be performed in a way that satisfies the familiar rules from the arithmetic of ordinary numbers. Concerning $\mathbb{F}_{2^b}$, a widely accepted approach is to represent the elements as polynomials of degree strictly less than b over $\mathbb{F}_2$. Operations are then performed modulo R where R is an irreducible polynomial of degree b over $\mathbb{F}_2$. For example, the field $\mathbb{F}_8$ can be interpreted as the binary polynomials modulo $1 + x + x^3$. This way we can consider $\mathbb{F}_8$ as the set of binary polynomials of degree at most 2 (indeed there are 8 such polynomials). Addition is the usual binary polynomial addition. For example,

$$(1 + x + x^2) + (1 + x^2) = x.$$

Multiplication is the usual polynomial multiplication, followed by reduction if necessary (modulo $1 + x + x^3$). By reduction we mean replacing x^3 by $x + 1$ as long as it is possible. For example,

$$(1+x+x^2)x^2 = x^2+x^3+x^4 = x^2+(x+1)+x(x+1) = x^2+x+1+x^2+x = 1.$$

In this representation x is a *primitive element*, indeed 7 is the smallest positive integer exponent m for which $x^m = 1$.

Lemma 3.3. *Let $\mathfrak{B} := \lceil \log_2(n+1) \rceil$ and $\mathfrak{B} > 2$. Then a series of $n \leq 2^{\mathfrak{B}} - 1$ nonzero binary codes $\mathbf{r}_1, \mathbf{r}_2, \ldots, \mathbf{r}_n$ of length $\mathfrak{B}$ can be generated in polynomial time to satisfy properties (A1), (A2) and (A3).*

Proof. With $\mathfrak{B} := \lceil \log_2(n+1) \rceil$, $q = 2^{\mathfrak{B}}$ is the smallest power of 2 which is greater than n. Following the widely used technique in classical error correcting codes, our code vectors will be vectors from a linear space over the two-element field $\mathbb{F}_2$. We shall consider the finite (Galois) field $\mathbb{F}_q$ with q elements.

According to Theorem 2.5 in [21], $\mathbb{F}_q$ always exists and it forms a vector space of dimension $\mathfrak{B}$ over its subfield $\mathbb{F}_2$. This way we can identify $\mathbb{F}_q$ with bit vectors of length $\mathfrak{B}$, where the all zero vector corresponds to the 0 element of $\mathbb{F}_q$. In particular, nonzero vectors correspond to the nonzero elements of the field. Also, according to Theorem 2.8 in [21], $\mathbb{F}_q$ contains a primitive element α, which is a nonzero element such that all the powers $\alpha = \alpha^1, \alpha^2, \ldots, \alpha^{q-1}$ are pairwise different. See also Table 3.4 where the elements and the related codes are listed for $q = 8$ ($\mathfrak{B} = 3$).

Table 3.4 The nonzero elements of $\mathbb{F}_8$ as binary polynomials modulo $1 + x + x^3$

Exponential	Polynomial	Code
α^0	1	$\mathbf{r}_1 = 100$
α^1	x	$\mathbf{r}_2 = 010$
α^2	x^2	$\mathbf{r}_3 = 001$
α^3	$x^3 = 1 + x \mod 1 + x + x^3$	$\mathbf{r}_4 = 110$
α^4	$x + x^2$	$\mathbf{r}_5 = 011$
α^5	$x \cdot (x + x^2) = 1 + x + x^2 \mod 1 + x + x^3$	$\mathbf{r}_6 = 111$
α^6	$x \cdot (1 + x + x^2) = 1 + x^2 \mod 1 + x + x^3$	$\mathbf{r}_7 = 101$

Finding a primitive element in $\mathbb{F}_q$ can be done in polynomial time with exhaustive search, because any nonzero element α can be verified for being a primitive element by raising to a power and checking if the power equals to 1 with an exponent less than $q - 1$.

We now set $\mathbf{r}_i$ to be the (bit vector of the) element α^{i-1}. Condition (A1) is satisfied as $n \leq 2^\mathfrak{B} - 1$.

Suppose now that (A2) fails. Then there must exist $0 \leq i < j < n - 1$ such that $\alpha^i \oplus \alpha^{i+1} = \alpha^j \oplus \alpha^{j+1}$ holds in $\mathbb{F}_q$. But then we have $\alpha^i(1 \oplus \alpha) = \alpha^j(1 \oplus \alpha)$ which (using that $\mathfrak{B} > 1$ and hence that $1 \oplus \alpha$ is not 0) would imply that $\alpha^i = \alpha^j$ and $\alpha^{j-i} = 1$, contradicting the fact that α is a primitive element.

To establish (A3), we note that (assuming $\mathfrak{B} > 2$) α^{-1} and α^{n-1} span a subspace of dimension at most 2 of $\mathbb{F}_q$ over $\mathbb{F}_2$, hence we can select the basis of $\mathbb{F}_q$ so that both elements have 0 coordinates with respect to the first basis vector.

3.2.4.2 The Proposed Construction for Chocolate Bar Graphs

The bit vectors attached to the edges of $C_n(V, E)$ define $\mathfrak{B}$ trails $T_1, \ldots, T_\mathfrak{B}$. These are actually simple paths in $C_n(V, E)$ from $v_{1,0}$ to $v_{0,n}$. As a result, $\mathfrak{B}$ m-trails from $v_{1,0}$ to $v_{0,n}$ are formed in C_n, each corresponding to one bit position of the vectors. An example with $n = 6$, where the resultant 3 m-trails by the construction are shown in Fig. 3.7b–d.

In addition to the above mentioned bm-trails, we need to add two more m-trails. This is exemplified in Fig. 3.7e, f. Let the two m-trails correspond to $T_{\mathfrak{B}+1}$ and $T_{\mathfrak{B}+2}$, respectively, where $T_{\mathfrak{B}+1}$ is composed of the links $(v_{1,0}, v_{0,0})$, $(v_{1,n}, v_{0,n})$ and the path consisting of all the links $(v_{1,i}, v_{1,i+1})$ $i = 0, \ldots n - 1$, while $T_{\mathfrak{B}+2}$ is composed of the links $(v_{1,0}, v_{0,0})$, $(v_{1,n}, v_{0,n})$ along with the path consisting of all the links $(v_{0,i}, v_{0,i+1})$, $i = 0, \ldots n - 1$. As a result, $T_{\mathfrak{B}+1}$ and $T_{\mathfrak{B}+2}$ can identify whether a failed link was a horizontal or vertical link, and whether the link was $(v_{1,0}, v_{0,0})$ or $(v_{1,n}, v_{0,n})$. Note that each T_j $j = 1, \ldots, \mathfrak{B} + 2$ forms a simple path.

3.2 UFL for Single Failures

3.2.4.3 Correctness of the Constructed Solution

We will show in the following paragraphs that the set of m-trails $T_1, \ldots, T_{\mathfrak{B}+2}$ are able to localize any single-link failure in chocolate bar $C_n(V, E)$. Obviously, T_1, $T_{\mathfrak{B}+1}$, and $T_{\mathfrak{B}+2}$ can unambiguously localize any failed link among $(v_{1,0}, v_{0,0})$ and $(v_{1,n}, v_{0,n})$ because the faulty link can be one of $(v_{1,0}, v_{0,0})$ or $(v_{1,n}, v_{0,n})$ if and only if both $T_{\mathfrak{B}+1}$ and $T_{\mathfrak{B}+2}$ are faulty. If both $T_{\mathfrak{B}+1}$ and $T_{\mathfrak{B}+2}$ alarm (i.e., report failure), the status of T_1 can be used to determine which of the two links $(v_{1,0}, v_{0,0})$ or $(v_{1,n}, v_{0,n})$ is at fault according to (A3).

For the other links, the statuses of $T_{\mathfrak{B}+1}$ and $T_{\mathfrak{B}+2}$ can be used to determine whether the faulty link is in the group of lower links, the group of upper links, or the group of middle links. With (A1), the links in the first two groups are pairwise different, while (A2) implies that the codes in the group of middle links are pairwise different. Therefore, all the links in each of the three groups are distinguishable such that UFL is possible within each group, and hence in $C_n(V, E)$.

3.2.4.4 The Number of M-Trails in the Construction

Since the chocolate bar graph has $3n + 1$ links, we have

$$\mathfrak{B} = \lceil \log_2(\frac{|E|-1}{3} + 1) \rceil \leq \lceil -1.58 + \log_2(|E| + 2) \rceil.$$

As a result the construction requires at most $b = \mathfrak{B} + 2 \leq \lceil 0.42 + \log_2(|E| + 2) \rceil$ m-trails.

3.2.5 An Essentially Optimal BM-Trail Solution for 2D Grid Topologies

In this section, the construction for the chocolate bar graphs is generalized for 2D rectangular grids. An $(m + 1)$-by-$(n + 1)$ grid graph is denoted as $S_{m,n}$, whose vertices are denoted as $v_{i,j}$ for $0 \leq i \leq m$ and $0 \leq j \leq n$. The vertical links of $S_{m,n}$ are $(v_{i,j}, v_{i+1,j})$ for $0 \leq i < m$ and $0 \leq j \leq n$. Analogously, the horizontal links of $S_{m,n}$ are $(v_{i,j}, v_{i,j+1})$ for $0 \leq i \leq m$ and $0 \leq j < n$.

Theorem 3.6. *A 2D rectangular grid graph $S_{m,n}(V, E)$ can be covered with $5 + \lceil \log_2(|E| + 1) \rceil$ bm-trails to achieve UFL, for $m, n \geq 1$.*

Proof. We shall have two monitoring sets of bm-trails. The first set has size $b_1 = \lceil \log_2(m + 1) \rceil + 2$, while the second has size $b_2 = \lceil \log_2(n + 1) \rceil + 2$. Informally speaking, the first set gives the horizontal position of a failed link, while the other gives the vertical coordinate. This will be sufficient to locate the failed link unambiguously. In total, we shall have no more than $\mathscr{B} = b_1 + b_2$ monitoring bm-trails.

3.2.5.1 The Proposed Construction

We are going to extend the bm-trails T_i ($i = 1, \ldots, b_1$) from the chocolate bar graph C_n of the first two rows to the whole square grid $S_{m,n}$. We do it step by step as follows: first we reflect the bm-trail T_i with respect to the line connecting $v_{1,0}$ to $v_{1,n}$, such that T_i is extended to the second chocolate bar defined by the vertices $v_{1,j}$ and $v_{2,j}$, for $j = 0, \ldots, n$. The second chocolate bar is extended analogously by reflection to the third chocolate bar, defined by $v_{2,j}$ and $v_{3,j}$, and so on. This reflection process is repeated until the whole $S_{m,n}$ is covered, where the 2D rectangular grid is treated as a series of chocolate bar graphs of C_n. As shown in Fig. 3.8a, at every second line the chocolate bar graph is upside down, and the ith chocolate bar graph C_n consists of vertices $v_{i,0}, \ldots, v_{i,n}$ and $v_{i+1,0}, \ldots, v_{i+1,n}$, where $i = 0, \ldots, m-1$. By applying the reflection process for all the bm-trails T_i ($i = 1, \ldots, b_1$), we will obtain b_1 bm-trails.

It is clear that the result of the reflection process must be a connected subgraph, thereby its eligibility as a bm-trail is ensured.

With the whole situation transposed, exactly the same method is applied to specify the vertical position i of the faulty link e. For the remaining b_2 bm-trails of the rectangular grid $S_{m,n}$ we start out with the vertically placed chocolate bar C_m^T at the left end of the grid (see Fig. 3.8b) and extend the b_2 bm-trails of this C_m^T to the whole grid with the mirror-reflection procedure employed before, nonetheless from left to right in order to extend the bm-trails to all the vertical chocolate bars in the grid. By doing this b_2 bm-trails can be obtained.

With the b_1 and b_2 bm-trails, we completed the construction.

3.2.5.2 Correctness of the Constructed BM-Trail Solution

In case of a single failure T_{b_1-1}, T_{b_1}, $T_{\mathscr{B}-1}$, and $T_{\mathscr{B}}$ (see also Fig. 3.8e–h) can identify whether a horizontal or a vertical link has failed and if the link is on the left or right border of the rectangular grid (it is on the first or last row/column). Since the failed link belongs to at least one of the horizontal chocolate bar graph C_n, the corresponding $b_1 - 2$ bm-trails can identify the column of the failed link. Similarly, the failed link belongs to at least one of the vertical chocolate bar graph C_m^T, the corresponding $b_2 - 2$ bm-trails can identify the row of the failed link. As a result, it is known if the link is horizontal or vertical, and its column and row thus can be localized.

3.2 UFL for Single Failures   59

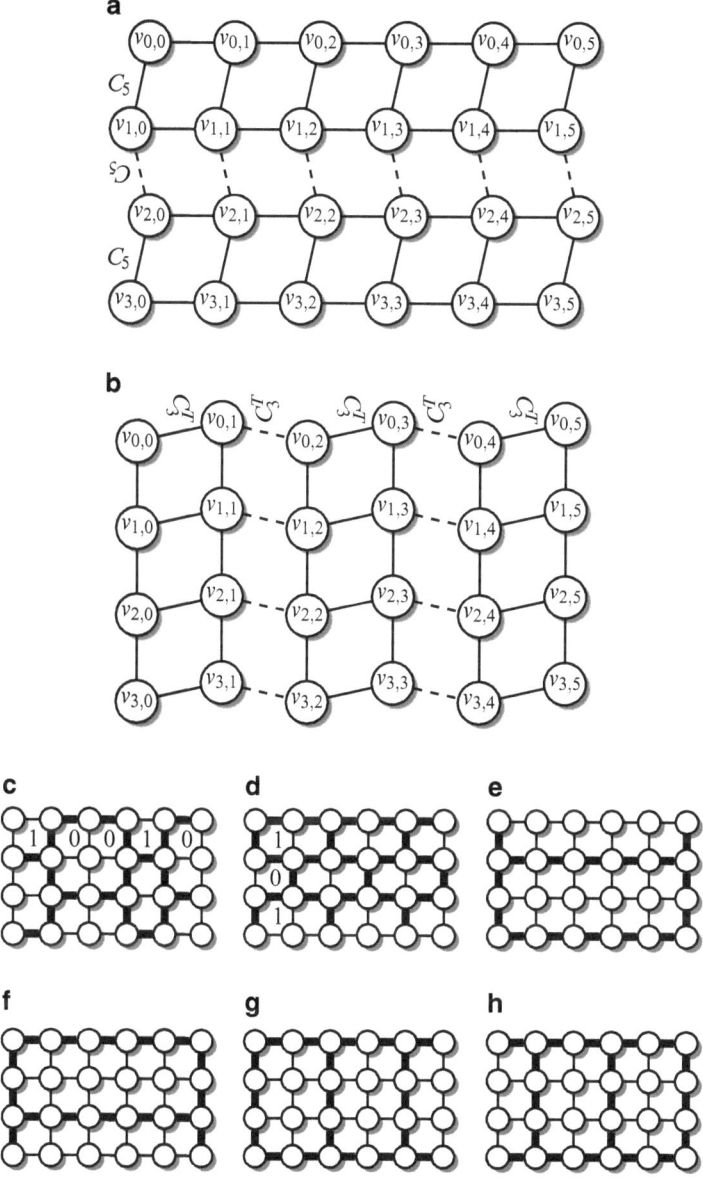

Fig. 3.8 An example of a 2D lattice graph of size 3×5. (**a**) $S_{3,5}$ decomposed into chocolate bar graphs in horizontal way. (**b**) $S_{3,5}$ decomposed into chocolate bars in vertical way. (**c**) An example of T_i for $i = 1, \ldots, b_1 - 2$. (**d**) An example of T_i for $i = b_1 + 1, \ldots, \mathscr{B} - 2$. (**e**) The links of T_{b_1-1}. (**f**) The links of T_{b_1}. (**g**) The links of $T_{\mathscr{B}-1}$. (**h**) The links of $T_{\mathscr{B}}$

3.2.5.3 The Number of BM-Trails in the Construction

Since $S_{m,n}$ has totally $|E| = 2 \cdot m \cdot n + n + m$ links, the number of bm-trails is:

$$\begin{aligned}
\mathscr{B} &= \lceil \log_2(m+1) \rceil + \lceil \log_2(n+1) \rceil + 4 \\
&\leq \lceil 1 + \log_2(m+1) + \log_2(n+1) \rceil + 4 \\
&= \lceil \log_2 2 + \log_2(m+1) + \log_2(n+1) \rceil + 4 \\
&= \lceil \log_2(2 \cdot (m+1) \cdot (n+1)) \rceil + 4 \\
&= \lceil \log_2(2mn + 2n + 2m + 2) \rceil + 4 \\
&= 4 + \lceil \log_2(2|E| - 2mn + 2) \rceil \\
&< 4 + \lceil \log_2(2|E| + 2) \rceil = 5 + \lceil \log_2(|E| + 1) \rceil \quad (3.8)
\end{aligned}$$

for $m, n \geq 1$. Note that the first inequality holds because of the general inequality $\lceil A \rceil + \lceil B \rceil \leq \lceil A + B \rceil + 1$, while the second follows from $m \cdot n > 0$.

A somewhat more complicated construction for the chocolate bar graph gives here a slightly better construction with only $3 + \log_2(|E| + 1)$ trails. We omit the details.

More generally, a similar construction can be used to cover a cubic graph of any dimension for single-link UFL with $O(1) + \log_2 |E|$ bm-trails. In this case the alarm code is divided into three parts, and each of them corresponds to a chocolate bar graph.

3.2.5.4 Chocolate Bar Graph as a Benchmark

Simulation is conducted on chocolate bar topologies with different sizes, aiming to examine the performance of a number of previously reported heuristic algorithms. We will show that the heuristics perform badly in chocolate bar topologies, which nonetheless can be optimally solved with the proposed construction by a very fast algorithm.

The reported heuristic algorithms for solving the MAP in chocolate bars are as follows.

1. *RCA–RCS* heuristic of Sect. 3.2.7.
2. *MTA* heuristic by Zhao et al. [53], which is a deterministic approach that builds the m-trails in parallel until UFL is achieved.
3. *RNH* heuristic by Zhao et al. [54] which is a randomized version of the *MTA* heuristic.
4. *Cycle Accumulation (CA)*: a generic approach by employing Dijkstra's algorithm to distinguish each pair of links [2].

3.2 UFL for Single Failures

We consider the three schemes: RCA–RCS, MTA, and RNH in chocolate bar graphs C_{20} and C_{60}. A high-performance server with 3 GHz Intel Xeon CPU 5160 was used in the simulation. The result of the proposed 2D grid construction in Sect. 3.2.4.2 is $\lceil \log_2(n+1) \rceil + 2$, which yields the minimum number of bm-trails for UFL as 7, and 8 for C_{20} and C_{60} with $|E| = 61$, and 181, respectively. We are interested in the difference between the result by each considered heuristic and the one obtained via the proposed construction. It is important to note that RCA–RCS and RNH are both randomized approaches where longer computation time guarantees better performance (or smaller numbers of bm-trails). Therefore, we are further interested to see how long does it take to the two schemes to converge close to the optimal solution, where the computation time describes the efficiency (and inefficiency) of the two schemes in the considered scenarios.

Figure 3.9 demonstrates the comparison results. Clearly, both RCA–RCS and RNH show better solution quality by granting longer running time, while MTA is a deterministic algorithm that iteratively finds the longest segment as the next m-trail. Since all the three topologies are very sparse whose diameters are much longer than the average nodal degree, MTA needs 6 and 15 more bm-trails in C_{20} than the optimum (i.e., 7), respectively, as shown in Fig. 3.9a, b. On the other hand, both RCA–RCS and RNH are seen to converge very slowly as the network has a larger diameter. As shown in Fig. 3.9a, RCA–RCS needs to take over 400 seconds to achieve a solution with one more trail than the construction for C_{20}, but over 800 and 1,000 s to approximate the construction within 10 trails for C_{60}, respectively, as shown in Fig. 3.9b, c.

Note that CA requires 93 bm-trails which are not shown in the figures since it is out of the range. We do not show RNH in Fig. 3.9b, c, and MTA in Fig. 3.9c, because they were not solvable in the topologies C_{20} by running out of 2GB memory (which is the computation specification for the simulation). This is mainly because both methods are designed for networks when the average nodal degree is not much smaller than the diameter of the network. Therefore, the two schemes had a large amount of candidate segments which drained the memory usage.

3.2.6 Optimal BM-Trail Solution for Circulant graphs

A circulant graph $C_n(1,2)$ has nodes $V = \{v_0, v_1, \ldots, v_{n-1}\}$ with each node v_j adjacent to $v_{[j+1 \bmod n]}$ and $v_{[j+2 \bmod n]}$. As an example, a circulant $C_9(1,2)$ is given in Fig. 3.10. Circulant graphs are considered to have similar properties to some practical carrier topologies.

Theorem 3.7. *The circulant graph* $G = C_n(1,2)$ *can be covered with* $b = \lceil \log_2(2n+1) \rceil$ *bm-trails for NL UFL, where each bm-trail is a spanning subgraph of* G, *and the bandwidth cost is* $\|\mathcal{T}\| = b \cdot n + 1$.

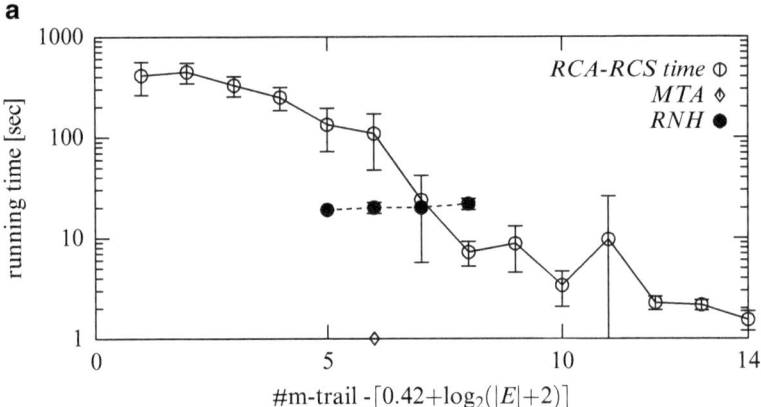

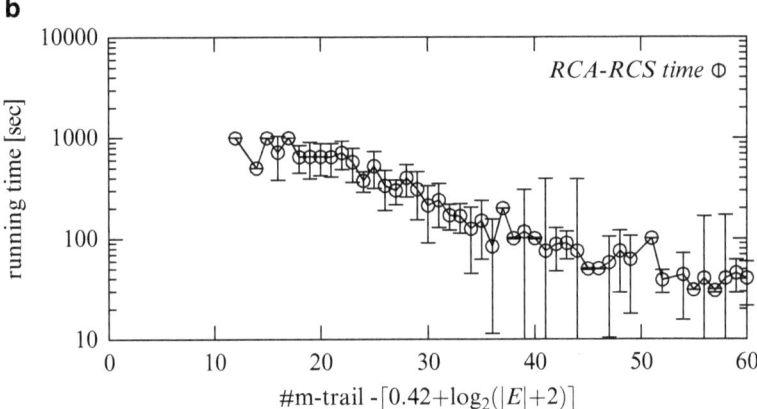

Fig. 3.9 Simulation results on C_{20} and C_{60} for the number of bm-trails. (**a**) C_{20}, (**b**) C_{60}

Fig. 3.10 An example of a circulant graph $G = C_9(1, 2)$

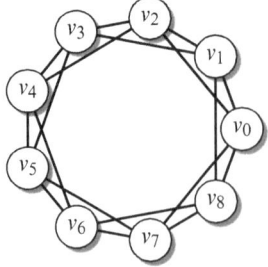

3.2 UFL for Single Failures

Proof. To prove the theorem, our approach is via a novel construction that generates a set of connected subgraphs of G as bm-trails, which can be proved to achieve UFL for any single-link failure.

3.2.6.1 The Proposed Construction

Let an alarm code of each link in G be b bits in length. The code $[11\ldots 1]^4$ is assigned to edge (v_0, v_1), while the other edges (v_i, v_{i+1}) are each assigned with an alarm code that is the binary representation of the value $i+1$ for $i = 1,\ldots,n-1$. In the special case $i = n-1$ the edge in question is (v_{n-1}, v_0) and the associated alarm code is the binary representation of n. Note that the first bit of these codes is always 0 and the rest is a nonzero bit vector. Moreover, these can be accomplished by using b bits only. Indeed, we have $2^b \geq 2n+1$ by assumption. This implies that $2^b \geq 2n+2$, giving that even the binary code of n fits in $b-1$ bits. Edge (v_i, v_{i+2}) is assigned with a code which is bitwise complement of the alarm code of (v_i, v_{i+1}) where $i \neq 0$. Thus, the first bit of these codes for (v_i, v_{i+2}) is 1. Besides, the complementary pair of codes $[00\ldots 01]$ and $[11\ldots 10]$ are not assigned to any edge. Finally, edge $(0, 2)$ is associated with the bit vector $[00\ldots 01]$.

The set of bm-trails is deployed in such a way that trail T_j traverses through all the edges with their jth bit value 1 while disjoint from any edges with the jth bit value 0.

3.2.6.2 Correctness of the Constructed BM-Trail Solution

It is clear that such an assignment generates a unique alarm code for each link. In the rest of the proof we show that for any $1 \leq j \leq b$, the subgraph T_j corresponding to links with their jth bit position of alarm codes 1 is connected and spans the whole vertex set. To make the proof easily presented, let us take each edge in G as directed counterclockwise, i.e., edge (v_i, v_{i+1}) is directed from v_i to v_{i+1}, and similarly (v_i, v_{i+2}) from v_i to v_{i+2}. It is sufficient to show that in T_j from every node v_i node 0 can be accessed. This is due to the following two facts: (1) the codes of (v_i, v_{i+1}) and (v_i, v_{i+2}) as $i \neq 0$ are bitwise complement to each other, thus T_j must connect from v_i to either v_{i+1} or v_{i+2}, this results every cycle to traverse the circulant graph counterclockwise through either node v_0 or v_1; and (2) when $i = 0$, edge (v_0, v_1) is in T_j for every j, since it has a code $[11\ldots 1]$.

---

[4] $x\ldots x$ denotes a code fragment with x in every bit position.

3.2.6.3 Number of BM-Trails in the Construction

To evaluate the total cost, the alarm codes for the directed edges leaving vertex v_0 contain exactly $b + 1$ values of 1 (i.e., edge (v_0, v_1) has b trails to traverse through, and edge (v_0, v_2) has one). For the other edges (v_i, v_{i+1}) and (v_i, v_{i+2}) with $i \neq 0$, we have b values of 1 altogether. This implies that the total number of 1s is $bn + 1$ as claimed.

The proposed construction is optimal in terms of the number of bm-trails as shown by the information theoretic lower bound $b \geq \lceil \log_2(|E|+1) \rceil$ and $|E| = 2n$.

3.2.7 The RCA–RCS Heuristic Approach for UFL

Surprisingly $\lceil \log_2(|E| + 1) \rceil$ m-trails are enough in typical network topologies. The proposed algorithm takes advantage of *random code assignment* (RCA) and *random code swapping* (RCS), aiming to overcome the difficulties coming from topology diversity. With RCA, it takes $|E|$ unique alarm codes which are randomly assigned to each link one after the other at the beginning and is kept in an alarm code table (ACT). This leads to $\lceil \log_2(|E| + 1) \rceil$ link sets. The algorithm then performs m-trail formation by examining the connectivity of each link set. There could be much more m-trails than $\lceil \log_2(|E| + 1) \rceil$ formed at the beginning. To improve the solution quality, RCS updates the ACT for each link set round by round, where a better structure of a link set is searched according to the cost function of Eq. (3.1). Note that RCS is performed independently (or locally) at each link set, where the codes of two links of different link sets can be swapped only if the swapping will not alter the connectivity of the other link sets. This is referred to as the *strong locality constraint* (SLC), which is an important feature of our design in making the algorithm simpler and running faster.

Figure 3.11 shows a flowchart of the proposed algorithm. At the beginning, an alarm code table (denote by $\underline{\underline{A}}$) is formed by randomly assigning each link with a unique alarm code as shown in Step (1). In Step (2), the cost of the current $\underline{\underline{A}}$ is evaluated by Eq. (3.1) (which will be further elaborated in Sect. 3.2.7.1). Next, a greedy cycle formed by Steps (2), (3), (4), (5), and (6) is initiated, where RCS is performed in Step (3) and (5) (which will be further detailed in Sect. 3.2.7.1). In every cycle, a new ACT (denoted as $\underline{\underline{A}}'$) is generated and the corresponding cost $\mathscr{C}'$ is evaluated in Step (2). If the cost of $\underline{\underline{A}}'$ (denoted as $\mathscr{C}'$) is smaller than (or equal to) that of the cost of previous $\underline{\underline{A}}$ (denoted $\mathscr{C}$) as in Step (2), the algorithm starts the next greedy cycle by replacing the old ACT with the new one (i.e., $\underline{\underline{A}} \leftarrow \underline{\underline{A}}'$) and performing RCS, denoted as $\underline{\underline{A}}' \leftarrow \Psi_{RCS}(\underline{\underline{A}}')$ in Step (3). In case the new ACT has a cost larger than that of the old one, the newly derived ACT is simply disregarded, and the next greedy cycle will perform RCS based on the old ACT again. Such a greedy cycle is iteratively performed until a given number of iterations of RCS have been done without getting a smaller cost at Step (4).

3.2 UFL for Single Failures

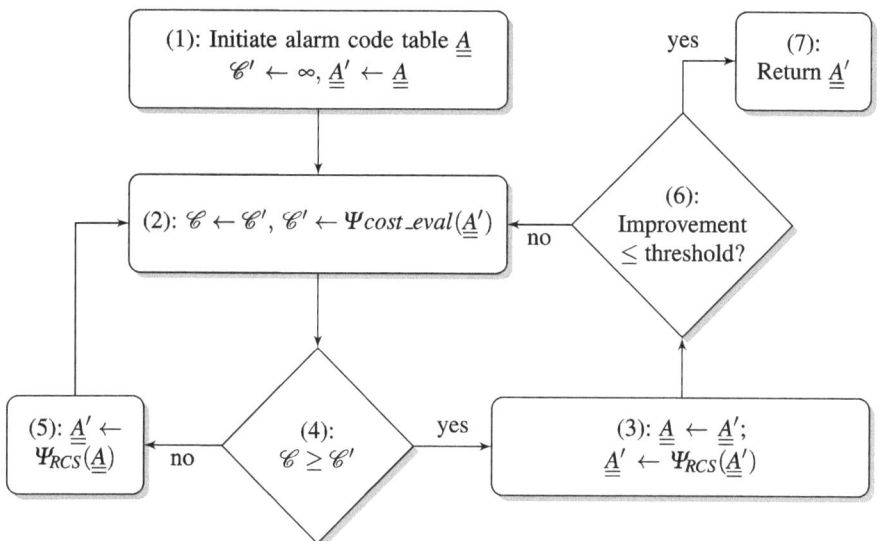

Fig. 3.11 The flowchart for the proposed heuristic algorithm

3.2.7.1 M-Trail Formation

This subsection introduces the basic idea of our m-trail formation mechanism, where Eq. (3.1) is used to evaluate $\underline{\underline{A}}$ in each greedy cycle in Fig. 3.11 such that the greedy cycle can possibly converge and yield a set of feasible m-trails with high quality.

Figure 3.12 explains Step (2) of Fig. 3.11 in detail through an example by considering a simple five-node topology. Initially, a 3-bit long alarm code is assigned to each link. The formation of the jth m-trail has to take all the links e with $a_{e,[j]} = 1$ (which belong to T_j) as shown in Fig. 3.12b–d. The ideal situation is that an $\underline{\underline{A}}$ with J bits yields exactly J link sets, which can form J valid m-trails. A link set forms an m-trail if all the links can be connected and traversed along a not necessarily simple path. In other words, the link set can have maximally two nodes with an odd nodal degree according to Euler's theorem. Checking this m-trail condition for a link set can be done by a breadth-first search (BFS) algorithm in linear time. A link set could be far from interconnected and could even yield multiple isolated fragments. If a link set cannot be shaped into a valid m-trail (e.g., link set 3 of Fig. 3.12d), it will possibly be constructed as a union of multiple m-trails or cycles according to the following Lemma.

Lemma 3.4. *A connected graph can be efficiently decomposed into (1) a single cycle, if every node has an even nodal degree; or (2) a number of #odd$(G)/2$ trails, where #odd(G) denotes the number of odd-degree nodes in the graph.*

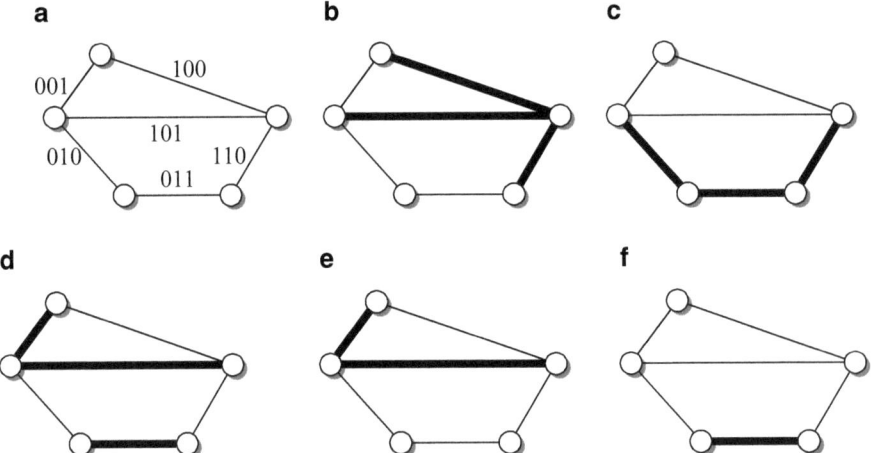

Fig. 3.12 An illustration of the cost evaluation method of an alarm code table of the 5 node example graph. (**a**) Alarm code of each link, (**b**) Link set T_1, (**c**) Link set T_2, (**d**) Link sets T_3 and T_4, (**e**) M-Trail T_3, (**f**) M-Trail T_4

Proof. The lemma is a consequence of Euler cycle and path theory. In both cases (1) and case (2), the cycle and the trails can be formed in linear time with Fleury's algorithm.

The Lemma states that in case the ith isolated fragment of link set of bit j (denoted as $C_{i,j}$) has more than two odd nodes (denoted by $\#odd(C_{ij})$), then it can be decomposed into the $\#odd(C_{ij})/2$ m-trails.

In case at a specific bit position the links with "1" bit do not form a trail, we can always "separate" those links into several trails. And each trail is going to be a separate m-trail.

Random Code Swapping

The initial RCA may yield an unsatisfactory result that contains many isolated fragments and a large number of odd-degree nodes. This subsection describes the proposed RCS mechanism for shaping the links of a link set into one or a number of m-trails while still meeting the overall UFL requirement. Let L_j be the set of links having "1" at the jth bit position of their current alarm code. The key idea of the proposed RCS mechanism is the SLC which governs the swapping mechanism in each link set. It means that the alarm code of a specific link in L_j can be swapped with that of a link not in L_j if all the other link sets are not affected due to the swapping. The necessary condition for meeting the SLC is that the alarm codes of two links in L_j are bitwise identical except for a single bit at position j. Such a code pair is called a *code pair of L_j*, and the two links corresponding to the code

3.2 UFL for Single Failures

pair form a *bitwise link-pair of L_j*. For example, [1011011] and [1010011] form a code pair of L_4, and the corresponding links form a bitwise link-pair of L_4. Thus, swapping alarm codes of the two links meet the SLC due to the local influence on L_4. With the SLC, the RCS on a link set can be performed independently from the others. This mechanism allows easy implementation and provides high efficiency.

Note that for L_j, some links may not have a bitwise link-pair due to two reasons: (1) its code pair of L_j is all 0's, which does not correspond to any failure state. For example, [010000] is a code pair of [000000] of L_2, but there is not a link corresponding to the alarm code [000000]. (2) The code pair of L_j of the link was not assigned to any link. In this case, the unassigned code can be freely used by the link without violating the SLC.

In summary, RCS is performed on each link set by randomly swapping alarm codes of all bitwise link-pairs of the link set, in order to help interconnecting isolated trail fragments and reducing the number of odd-degree nodes in the link set iteratively in each greedy cycle. The prototype of the proposed algorithm can be found in [35].

An Example on RCS Algorithm

We provide an example here to show how RCS is performed. A 26-node network of US cities considered with 42 links. Initially 6 bit long unique random codes were assigned to the links. Figure 3.13a shows the link set assigned to the lowest bit (i.e., the 6th). Except for the link between Denver and Kansas City that was assigned with an alarm code [0000001], all the other links either have a bitwise link-pair in L_6, or are don't-care links of L_6. Figure 3.13a shows each link-pair at L_6 by an arrow. For example, (Atlanta, Charlotte) and (Indianapolis, Cleveland) are bitwise link-pairs of L_6. It can be easily seen that swapping the two links will interconnect two isolated fragments and reduce the number of odd-degree nodes by two, which leads to a saving of an m-trail. Similarly, swapping link (Salt Lake City, Denver) with link (Houston, New Orleans) will not increase the total cost of the ACT. While in the subsequent greedy cycle, swapping link (Las Vegas, El Paso) with link (El Paso, Houston) would further reduce the number of m-trails through the RCS and thus possibly reduce the total cost. With more greedy steps on those link pairs and don't-care links of L_6, it is possible to form a single m-trail corresponding to the 6th bit in the ACT while keeping all other link sets intact. By iterating the greedy process for each bit in the ACT, the algorithm can guarantee to obtain an m-trail solution for each bit in the ACT. The solution quality will depend on the success rate threshold defined in Step (6) in Fig. 3.11. The effectiveness and efficiency of the proposed algorithm will be further demonstrated in Sect. 3.4.1.

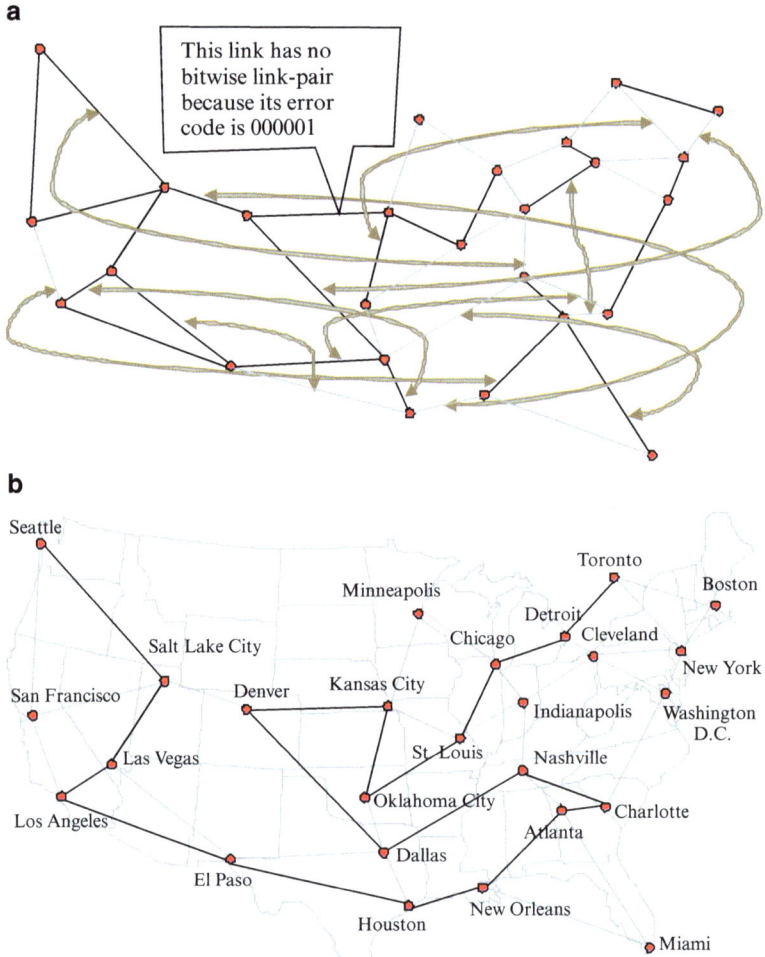

Fig. 3.13 An example link set. (**a**) The bitwise link pairs. (**b**) After greedy random code swapping

3.3 UFL for Multiple Failures

3.3.1 Problem Definition and Background

The alarm code of a multi-link SRLG is the *bitwise OR* of the alarm codes of all links in the SRLG (see also Fig. 3.14 as an example solution for SmallNet for dual-link failures). This corresponds to the fact that a monitor alarms if the corresponding monitoring lightpath traverses through any link in the SRLG that is hit by the failure event. Thus, each SRLG $z \in \mathscr{Z}$ can be assigned with an alarm code, denoted by $a_{\{z\}}$, as follows:

3.3 UFL for Multiple Failures

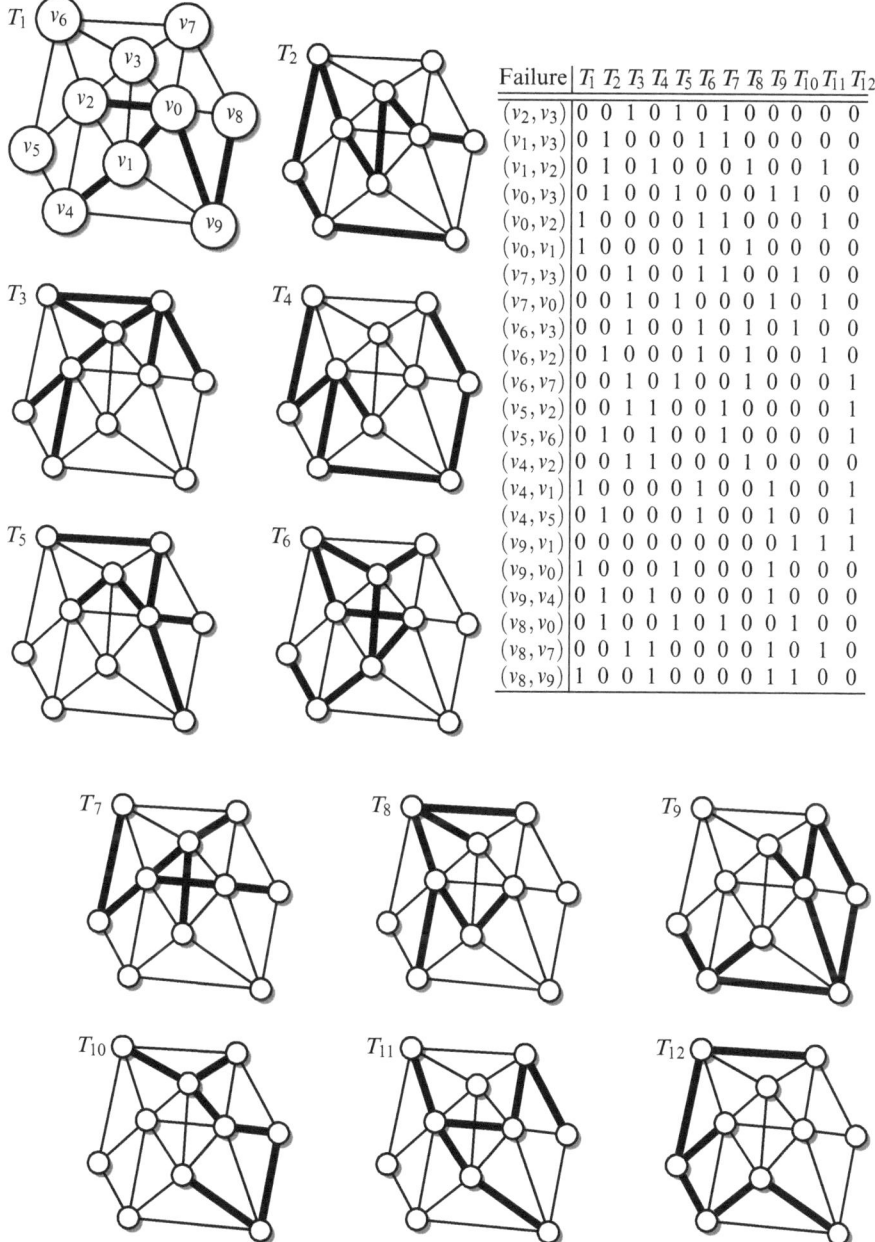

Failure	T_1	T_2	T_3	T_4	T_5	T_6	T_7	T_8	T_9	T_{10}	T_{11}	T_{12}
(v_2, v_3)	0	0	1	0	1	0	1	0	0	0	0	0
(v_1, v_3)	0	1	0	0	0	1	1	0	0	0	0	0
(v_1, v_2)	0	1	0	1	0	0	0	1	0	0	1	0
(v_0, v_3)	0	1	0	0	1	0	0	0	1	1	0	0
(v_0, v_2)	1	0	0	0	0	1	1	0	0	0	1	0
(v_0, v_1)	1	0	0	0	0	1	0	1	0	0	0	0
(v_7, v_3)	0	0	1	0	0	1	1	0	0	1	0	0
(v_7, v_0)	0	0	1	0	1	0	0	0	1	0	1	0
(v_6, v_3)	0	0	1	0	0	1	0	1	0	1	0	0
(v_6, v_2)	0	1	0	0	0	1	0	1	0	0	1	0
(v_6, v_7)	0	0	1	0	1	0	0	1	0	0	0	1
(v_5, v_2)	0	0	1	1	0	0	1	0	0	0	0	1
(v_5, v_6)	0	1	0	1	0	0	1	0	0	0	0	1
(v_4, v_2)	0	0	1	1	0	0	0	1	0	0	0	0
(v_4, v_1)	1	0	0	0	0	1	0	0	1	0	0	1
(v_4, v_5)	0	1	0	0	0	0	0	1	0	0	0	1
(v_9, v_1)	0	0	0	0	0	0	0	0	0	1	1	1
(v_9, v_0)	1	0	0	0	1	0	0	0	1	0	0	0
(v_9, v_4)	0	1	0	1	0	0	0	0	1	0	0	0
(v_8, v_0)	0	1	0	0	1	0	1	0	0	1	0	0
(v_8, v_7)	0	0	1	1	0	0	0	0	1	0	1	0
(v_8, v_9)	1	0	0	1	0	0	0	0	1	1	0	0

Fig. 3.14 Unambiguous failure localization (UFL) based on hm-trails for dual failures. The solution has $b = 12$ and $\|\mathcal{T}\| = 81$

$$a_{\{z\}} = \bigvee_{e \in z} a_e,$$

where $\vee$ denotes the *bitwise OR* operator and a_e is the alarm code of link e. The *UFL constraint* in this general case requires every alarm code of each SRLG to be unique. Thus, an effective monitoring structure allocation method must satisfy the following two requirements, either in a single step or one after the other:

(R1): Every SRLG should be uniquely coded.
(R2): Each monitoring structure must be an eligible fragment of network topology in which a lightpath can travel along from the transmitter to the receiver.

Note that in addition to (R1) and (R2), there could be some other constraints due to specific user design premises, such as the length limitation due to the deployment of optical generators/retransmitters, the locations of monitoring nodes [2, 47], and the use of working lightpaths (i.e., live connections) for failure state correlation [2, 33].

An ILP can be developed that formulates both (R1) and (R2) in a single step [44–46]. In particular, [44] is the first study that suggested to using freely routed open-loop undirected supervisory lightpaths (m-trail) for single-link SRLG failure localization. All the three studies formulated the supervisory lightpath allocation problem into ILPs, which are unfortunately subject to intolerably long computational time even in very small topologies. Thus people have turned to the design of heuristics in solving the problem. The previously reported solutions can be divided into two categories according to their design principles. The first one manipulates an accumulation mechanism such that (R2) is ensured at the beginning, while the goal of the heuristics is to satisfy (R1) [2, 3, 18]. In the second design category, (R1) is intrinsically ensured at the beginning while leaving (R2) as a goal [40].

The bm-trail formation problem is a structured variant of the combinatorial group testing (CGT) process [13, 19]. In [18] the problem was taken as *combinatorial group testing on graphs*. The idea of group testing dates back to World War II when millions of blood samples were analyzed to detect syphilis in US military. In order to reduce the number of tests it was suggested to pool the blood samples. From algorithmic aspects there are two significant differences between the tasks of pooling blood samples and monitoring of a group of links in a graph: (1) the blood samples can be pooled arbitrarily, while the monitored links must be links of a connected sub-graph and in a valid shape, (2) the monitors are always pre-configured, and the probing is performed simultaneously without knowing the result of other tests (non-adaptive CGT).

The primary goal of any CGT algorithm is to identify defective items among a given set of items through as few tests as possible. In this case the set of items are the links of a graph, the defective items are the failed links, and the tests are monitoring structures (e.g., bm-trails). A general CGT method was given by Hwang and Sós [19], while the shortest non-adaptive CGT codes for practical sized problems were developed by Eppstein et al. [14].

3.3 UFL for Multiple Failures

In [18], based on non-adaptive combinatorial group testing (CGT) constructions, the authors conducted a polynomial time construction for very densely connected networks, where the topology graph has $d + 1$ edge-disjoint spanning trees. In the construction each link is assigned with alarm code of $d \cdot j$ bits for achieving UFL of SRLGs with up to d links, where j is the length of a d-separable CGT codes for $|E|$ items. The method of assigning each link d CGT codes can fit well into theoretical analysis, but it can hardly be applied in most practical scenarios.

The studies in [2, 3] set their primary goal in minimizing the number of monitoring locations (MLs), i.e., the terminal nodes of m-trails, and payed little attention on minimizing the number of m-trails/cycles and the cover length. To identify the minimum number of ML that can localize the failure of SRLG with up to d links, all the d- and $(d + 1)$-connected subgraphs should be identified, and almost each subgraph needs an ML at an arbitrarily chosen node in the subgraph. With each ML determined, graph transformation is performed such that the MLs are merged into a supernode (denoted as m), and cycles are cumulatively added into the transformed graph one by one via Suurballe's [34] algorithm (see Sect. 7.1.1 for details). To distinguish two SRLGs w_1 and w_2, a cycle must be disjoint from w_2 while passing m and l, where l is a link randomly selected from $w_1 \setminus w_2$. In the worst case this leads to $O(|SRLG|^2)$ of cycles to distinguish all the SRLGs, where $|SRLG|$ is the number of SRLGs considered in the network. Thus, the worst time complexity is $O(|SRLG|^2 \cdot |V|^2)$, where $|V|$ is the number of nodes in network, and the term $O(|V|^2)$ corresponds to the complexity of Suurballe's algorithm. The computation complexity becomes $O(|E|^{2d} \cdot |V|^2)$ if every multiple failure with up to d links should be localized, where $|E|$ is the number of links.

The approach taken in [40] (see also Sect. 3.2.7) is the first study following the second design principle, where the code uniqueness of each link (as defined in (R1)) is first guaranteed, while an algorithm was given for the formation of each monitoring structure as an m-trail. A superb performance was witnessed in [40] by employing RCA and RCS for localizing any link failure. To be specific, the RCA algorithm forms the jth m-trail by randomly swapping a link code with its *bitwise code pair* at the jth position. For example, the codes [11010110] and [11000110] form a bitwise code pair at the 4th position. Note that such an RCS algorithm can only work when single-link failures are considered and it simply fails in presence of the code dependency among overlapped SRLGs, which is the most critical task to be addressed in this section.

This section follows the second design category in order to take advantage of the extremely flexible structure of bm-trails in solving the problem [6].

3.3.2 Computational Complexity of UFL for Multiple Failures

The previously mentioned ILP formulations in Sect. 3.3.1 were subject to intolerably long computational time even in very small topologies, which suggested that the MAP could be a hard problem. Here, we show that, in fact, for multiple failures the MAP is NP-complete. However, note that, for single-link failures it is an open question at the moment.

There are several studies investigating related problems, which show that handling multiple failures is a difficult problem. For example, the study [33] investigated the redundant alarm reduction problem based on a set of existing working lightpaths (*in-band monitoring*), which aimed to minimize the number of active alarms (or the number of monitors, which is the number of supervisory lightpaths in *out-of-band monitoring*) in the network while maintaining unambiguous single-link failure localization. It has been shown that the Redundant Monitor Deactivation Problem (RMDP) is NP-complete. Note that [33] relies on the existing lightpaths for UFL, without allocating any additional S-LPs for the purpose of monitoring.

Similarly to [33], the study in [27] studied the problem of localizing failures in transparent optical networks using active lightpaths for monitoring purposes. The problem of optimal allocation of monitoring devices was investigated, which was solved by using a transparent failure location algorithm (TFLA) for detecting both hard (e.g., cable cut) and soft (e.g., high Bit Error Rate (BER)) failures. Although no additional resources are needed for monitoring purposes, false positive or false negative could happen since some alarms correspond to multiple SRLGs, which results in ambiguity. The problem relies on the failure model in [30] in which the problem of uniquely diagnosed single faults and multiple faults are investigated. It has been shown in [30] that the localization of multiple link faults is an NP-complete problem if the faults propagate in the network.

The MAP can be formulated as follows:

Definition 3.1. The MAP instance: a network $G = (V, E)$, a $\mathscr{L}$ set of SRLGs, a positive integer $b_{max} < |E|$ representing a limit on the length of the link code. The MAP problem is to find a set of $b \leq b_{max}$ m-trails so as to unambiguously localize all SRLGs in $\mathscr{L}$; i.e., for any pair of SRLGs $e, f \in \mathscr{L}$, (1) there is an m-trail which traverses any link of e and disjoint from f or vice versa, and (2) all SRLGs are passed by at least one m-trail. Note that if an SRLG fails, all of its links assumed to be down.

With this SRLG model we tacitly assume that the failure of an SRLG implies the failure of all elements in the SRLG. A hardness result on the MAP problem is derived by means of a Karp reduction from the Hitting Set problem. The definition of the Hitting Set problem is as follows [16]:

Definition 3.2. Hitting Set problem instance: Collection C of subsets of a finite set S, positive integer $K < |S|$.

The hitting set problem is to find a subset $S' \subseteq S$ with $|S'| \leq K$ such that S' contains at least one element from each subset in C.

Theorem 3.8. *The MAP problem is NP-complete.*

Proof. The MAP problem is in NP, as we can efficiently verify whether a candidate solution unambiguously localizes all SRLGs with $\leq b_{max}$ m-trails.

Assume we are given an instance of the hitting set problem, that is, a finite set S with n elements $s_1, s_2, \ldots, s_n$, and a collection of subsets $C_1, C_2, \ldots, C_r$.

The polynomial time transformation is given as follows. We construct a graph with $n + r$ isolated edges, where edge e_i is assigned for each $s_i, i = 1, \ldots, n$ in the set, and isolated edge f_j is assigned for each subset $C_j, j = 1, \ldots, r$. Note that the graph consists of isolated edges, thus, all m-trails have length of a single edge. An SRLG z_j is defined for all $C_j = \{s_{j_1}, s_{j_2}, \ldots, s_{j_k}\}$ as SRLG $z_j = \{e_{j_1}, e_{j_2}, \ldots, e_{j_k}, f_j\}$, and SRLG $z_{r+j} = \{f_j\}$. In the rest of the proof we show that the above $2r$ SRLGs can be unambiguously localized with $\leq b = K + r$ m-trails, if and only if ($\Leftrightarrow$) the hitting set problem is solvable with $\leq K$ elements.

($\Leftarrow$) Suppose there exists a solution for the hitting set problem with K elements. In this case, MAP problem has a solution with $b = K + r$ m-trails. In the m-trail solution, all edges f_j for $j = 1, \ldots, r$ are covered by a single m-trail. Besides, the e_{j_k} edges corresponding to element s_{j_k} in the hitting set are also covered by a single m-trail. In this case, the SRLGs are unambiguously localized, because (2) all SRLGs are covered with at least one m-trail; and (1) the m-trail on f_i passes through SRLG z_i and SRLG z_{i+r}, but none of the other SRLGs. The two SRLGs can be distinguished if there exists an m-trail, which passes through SRLG z_i, but not SRLG z_{i+r}. Such m-trail exists, as SRLG z_i is passed by the m-trail on the single edge e_{j_k}, which corresponds to the s_{j_k} element in the hitting set (and at least one such element exists).

($\Rightarrow$) Finally, we show how to convert an m-trail solution with $b \leq K + r$ m-trails into a K element solution of the corresponding hitting set problem. The first observation is that the MAP solution has r m-trails with single links of f_j for $j = 1, \ldots, r$ for cover (2) and unambiguously localize SRLG $z_i, i = r+1, \ldots, 2r$. The second observation is that at least one edge e_i from each SRLG $z_j, j = 1, \ldots, r$ is covered by an m-trail in order to distinguish the failure of SRLG z_i and SRLG z_{i+r}. The edges e_i for $i = 1, \ldots, n$ are passed by $K = b - r$ single-hop m-trails and the K elements in S corresponding to these m-trails form a hitting set on the subsets $C_j, j = 1, \ldots, r$. Thus, the MAP is NP-complete.

As a consequence of this result, finding a fast algorithm that solves an arbitrary input of the MAP problem is unlikely (unless $P = NP$). Thus, in the next sections we introduce the ILP approach for solving the MAP problem, and enumerate the most relevant heuristic approaches.

3.3.3 Optimal UFL Solution for Multiple Failures

In this section the ILP constraints for the (R1) code assignment phase and the MAP (R2) are formulated for unambiguously localizing an SRLG list $\mathscr{L}$ [7, 8]. Furthermore, code assignment using CGT codes with special property is shown for the multiple failure case. Note that the number of m-trails is determined by the length of the CGT codes, which is not necessarily optimal. Thus, the formulations using CGT codes are not generally optimal; it only minimizes the bandwidth cost of the applied m-trail solution in Eq. (3.1). However, the application of CGT codes is the best known solution to handle the highly dependent alarm codes of a given multiple failure list $\mathscr{L}$. An example is given, where a suite of rules is provided in order to guide the convergence of a heuristic approach (called Greedy Code Swapping, GCS^d, see Sect. 3.3.7) of the problem solution for better quality when all the SRLGs with up to d links are considered. However, when the SRLGs have significantly different sizes, then the CGT approach may not be efficient and another approach is required.

An ILP was formulated for single-link failures in [43]. The following formulation is the extended version of the previous formulation for multiple failures, thus, it has the ability to handle the dependency between the link codes and SRLG alarm codes, which were not present in the original formulation. The notations used in the ILP are summarized in Table 3.5.

Note that the undirected network is represented by directed arcs in the formulation (i.e., each edge is replaced by two anti-parallel arcs). Thus, the found m-trails are directed trails, but because of the symmetry of the model adding the directed m-trails into the reverse direction as well the optimal solution for the undirected network $G = (V, E)$ can be obtained. Our goal is to minimize the cost function presented in Eq. (3.1), namely the number of m-trails and the used wavelength channels for monitoring purposes, formulated in the following objective function:

$$\min \left\{ \gamma \cdot \sum_{j \in J} m_j + \sum_{j \in J} \sum_{(u,v) \in E} c_{u,v} \cdot e_{u,v}^j \right\}. \qquad (3.9)$$

As the number of m-trails corresponds to the hardware cost and signaling complexity in the network, we set $\gamma = 1000$ to emphasize the importance of the monitoring cost. The problem specific constraints are presented in the following subsections.

3.3.3.1 Code Assignment (R1) Constraints

The sparse- and dense-SRLG scenario requires two different approaches, as in the dense-SRLG case assigning CGT codes to the links is a sufficient and efficient method. However, for the sparse-SRLG case CGT codes result insufficiently long codes. Thus, in the sparse-SRLG case we have to be careful in the ILP about the alarm code uniqueness in the alarm code table between the SRLGs. The ILP formulation in [43] used the sparse code assignment method for single-link failure localization to ensure code uniqueness in $\underline{A}$. For sparse-SRLGs and multiple link failures the formulation has to be extended to handle code dependency between link codes and SRLG alarm codes in $\underline{\underline{A}}$ when sparse-SRLGs and multiple failures are considered. The extended constraints in the sparse-SRLG case are:

3.3 UFL for Multiple Failures

Table 3.5 The notations used in the ILP

	Parameters
$G = (V, A)$	The underlying directed graph with node set V and link set A corresponding to the anti-parallel arcs of the undirected edges in E.
$\mathscr{L}$	A given set of SRLGs $\mathscr{L} = \{g_1, g_2, \ldots g_d\}$. We assume that g_i always contains both representation of an undirected link.
$\mathscr{L}_s, \mathscr{L}_m$	The single-link and multiple link SRLGs in the SRLGs list $\mathscr{L} = \mathscr{L}_s \cup \mathscr{L}_m$. Note that, $g_i \in \mathscr{L}_s$ contains both anti-parallel arcs $g_i = \{(u, v), (v, u)\}$ of an undirected edge.
b_{max}	The maximum number of m-trails allowed in the solution.
j	m-trail index, where $j \in J = \{0, 1, \ldots, b_{max} - 1\}$.
$\mathscr{C}$	Set of Combinatorial Group Testing codes in the dense-SRLG code assignment.
k_i	A single code in the CGT code list $k_i \in \mathscr{C}$.
k_i^j	It is a constant with value 1 if code k_i has 1 in the jth position, and 0 otherwise.
γ	The cost ratio of a monitor to a supervisory wavelength.
$c_{u,v}$	The cost of a supervisory wavelength channel on link (u, v), and we assume $c_{u,v} = c_{v,u}$. If hop count is used as the cost metric, then $c_{u,v} = 1$. Otherwise, it can be a distance related or any other cost.
δ	A predefined small positive value ($|E|^{-1} \geq \delta > 0$). It is the minimum step of voltage increase along an m-trail.
β	A predefined small constant and $2^{-b_{max}} \geq \beta > 0$.
	Variables
m^j	Binary variable. It takes 1 if t_j is an m-trail, and 0 if the corresponding column does not contain an m-trail.
s_u^j	Binary variable. It takes 1 if node u is the source of m-trail t_j, and 0 otherwise.
d_u^j	Binary variable. It takes 1 if node u is the sink of m-trail t_j, and 0 otherwise.
z_u^j	Binary variable. It takes 1 if node u is traversed by m-trail t_j, and 0 otherwise.
$e_{u,v}^j$	Binary variable. It takes 1 if (u, v) is an on-trail vector of m-trail t_j, and 0 otherwise. In the formulation each m-trail is a (directed) path with a source and destination node. Note that for b-mtrails $e_{u,v}^j = e_{v,u}^j$, because they traverse every link in both directions.
$q_{u,v}^j$	Fractional variable. It is the voltage of vector (u, v) on m-trail t_j. It takes 0 if (u, v) is not an on-trail vector on t_j.
b_g^j	Binary variable, which takes 1 if any edge in SRLG g is an on-trail vector of m-trail t_j, and 0 otherwise (jth bit of the alarm code of SRLG g).
α_g	General integer variable, which is the decimal alarm code assigned to SRLG $g = \{(u, v), (v, u)\}$.
$f_{i_1}^{i_2}$	Binary variable. It ensures that two distinct SRLGs have different alarm codes.
$x_{u,v}^i$	Binary variable. It takes 1 if k_i is assigned to link (u, v), and 0 otherwise.

$$\forall g \in \mathscr{L} : \alpha_g = \sum_{j \in J} 2^j \cdot b_g^j, \qquad (3.10)$$

$$\forall g \in \mathscr{L} : \alpha_g \geq 1, \qquad (3.11)$$

$$\forall g_1, g_2 \in \mathscr{L}, g_1 \neq g_2 : \beta + \beta \cdot (\alpha_{g_1} - \alpha_{g_2}) \leq f_{g_1}^{g_2}, \qquad (3.12)$$

$$\forall g_1, g_2 \in \mathscr{L}, g_1 \neq g_2 : \beta + \beta \cdot (\alpha_{g_2} - \alpha_{g_1}) \leq 1 - f_{g_1}^{g_2}, \qquad (3.13)$$

$$\forall j \in J, \forall g_s \in \mathscr{L}_s : \sum_{(u,v) \in g_s} e_{u,v}^j = b_{g_s}^j, \qquad (3.14)$$

$$\forall j \in J, \forall g_m \in \mathscr{L}_m, \forall g_s \in g_m : \sum_{(u,v) \in g_s} e_{u,v}^j \leq b_{g_m}^j, \qquad (3.15)$$

$$\forall j \in J, \forall g_m \in \mathscr{L}_m : b_{g_m}^j \leq \sum_{g_s \in g_m} b_{g_s}^j. \qquad (3.16)$$

Constraint (3.10) assembles the binary alarm code bits and translates them into a decimal alarm code. Constraint (3.11) says that every SRLG must have a positive decimal alarm code, i.e., prevents zero alarm codes. Equations (3.12) and (3.13) ensure distinct decimal alarm codes for any pair of SRLGs, which is equivalent to ensuring unambiguous SRLG failure localization. Note that for two distinct SRLGs g_1 and g_2 $f_{g_1}^{g_2}$ takes 1 if $\alpha_{g_1} > \alpha_{g_2}$, and 0 if $\alpha_{g_1} < \alpha_{g_2}$. Finally, Eqs. (3.14)–(3.16) formulate the *bitwise OR* requirement, where a multiple-link SRLG can have "1" in the jth position of its alarm code only if any of the single-link SRLGs with a common edge with it has "1" in the jth position.

When multiple failures (dense-SRLGs) are considered, the constraints presented in Eqs. (3.10)–(3.16) have to be replaced with the ones introduced in Eqs. (3.17)–(3.21). In this case besides the dense SRLG list $\mathscr{L}$ the $\mathscr{C}$ list of CGT codes is given as the part of the input (e.g., generated with the bktrk in [14]), and the task is to find a minimal cost assignment of the codes to the undirected edges in E, where on each bit position a single trail is formed:

$$\forall i \in \mathscr{C}, \forall (u,v) \in E : x_{u,v}^i = x_{v,u}^i, \qquad (3.17)$$

$$\forall (u,v) \in E : \sum_{i \in \mathscr{C}} (x_{u,v}^i + x_{v,u}^i) = 2, \qquad (3.18)$$

$$\forall i \in \mathscr{C} : \sum_{(u,v) \in E} (x_{u,v}^i + x_{v,u}^i) \leq 2, \qquad (3.19)$$

3.3 UFL for Multiple Failures

$$\forall j \in J, \forall (u,v) \in E : 2 \cdot (e^j_{u,v} + e^j_{v,u}) \geq \sum_{i \in \mathscr{C}} k^j_i \cdot (x^i_{u,v} + x^i_{v,u}), \tag{3.20}$$

$$\forall j \in J, \forall (u,v) \in E : e^j_{u,v} + e^j_{v,u} \leq \sum_{i \in \mathscr{C}} k^j_i \cdot (x^i_{u,v} + x^i_{v,u}). \tag{3.21}$$

Constraints in Eqs. (3.17)–(3.19) establish a map between the directed links in the ILP model and the undirected links of the topology, and ensure that code assignment is injective on the undirected links. Equations (3.20) and (3.21) say that the monitoring structure formed on the jth position traverses a link if and only if its code has "1" in the jth position.

Note that this formulation can be used for the code assignment phase of single failures as well. In that case, instead of CGT codes $\mathscr{C}$ can contain arbitrary unique codes. Finally, in both the sparse- and dense-SRLG case Eq. (3.22) says that if $|\mathscr{L}|$ SRLGs need to be differentiated then at least $\lceil \log_2 (|\mathscr{L}| + 1) \rceil$ bits (or m-trails) are required:

$$\sum_{j \in J} m_j \geq \lceil \log_2 (|\mathscr{L}| + 1) \rceil, \tag{3.22}$$

3.3.3.2 Monitoring-Trail Formation (R2) Constraints

In the ILP formulation each m-trail is a directed path with a source and destination node. For monitoring structure formation, the following constraints are required:

$$\forall j \in J : \sum_{u \in V} s^j_u \leq 1, \tag{3.23}$$

$$\forall j \in J : \sum_{u \in V} d^j_u \leq 1, \tag{3.24}$$

$$\forall j \in J, \forall u \in V : \sum_{(u,v) \in E} (e^j_{u,v} - e^j_{v,u}) = s^j_u - d^j_u, \tag{3.25}$$

$$\forall j \in J, \forall (u,v) \in E : q^j_{u,v} \leq e^j_{u,v}, \tag{3.26}$$

$$\forall j \in J, \forall u \in V : d^j_u + \sum_{(u,v) \in E} (q^j_{u,v} - q^j_{v,u}) \geq \delta \cdot z^j_u, \tag{3.27}$$

$$\forall j \subset J, \forall (u,v) \in E : m_j \geq e^j_{uv}, \tag{3.28}$$

Constraints Eqs. (3.23) and (3.24) allow at most one source-destination pair for each m-trail. Constraint Eq. (3.25) expresses the flow conservation of T_j.

The voltage constraint [43] is employed to ensure that the solution is a single trail, as the aforementioned constraints could result in more than one trail (i.e., a path and a disjoint cycle). Equations (3.26) and (3.27) say that only an on-trail vector could have nonzero voltage and ensures that the solution will be a single m-trail (or a cycle). The voltages should increase along an m-trail, and with this constraint it is ensured that a single trail is formed. With these constraints, the obtained link set can be covered by a single supervisory lightpath. Constraint Eq. (3.28) identifies the m-trails on each bit position, as only those bit positions form an m-trail, which have at least 1 link covered.

$$\forall j \in J, \forall (u,v) \in E : e_{u,v}^j + e_{v,u}^j \leq 1, \tag{3.29}$$

$$\forall j \in J, \forall u \in V, \forall (u,v) \in E : e_{u,v}^j + e_{v,u}^j \leq z_u^j, \tag{3.30}$$

Constraint (3.29) ensures that an m-trail should traverse an edge only once. Note that an m-trail is modeled as a directed path in the ILP. Equation (3.30) stipulates that a node has an inbound or outbound vector, and the vector should be traversed by an m-trail.

3.3.3.3 Bidirectional Monitoring-Trail Formation (R2) Constraints

Bm-trails traverse every link in both directions and have the same node as source and destination. In this section we give the necessary modifications of the constraints introduced in Sect. 3.3.3.2 to cope with b-mtrails instead of m-trails. First, constraints Eqs. (3.14), (3.15), (3.24), (3.29), and (3.30) have to be removed from the ILP, and the following constraints should be added:

$$\forall j \in J, \forall (u,v) \in E : e_{u,v}^j = e_{v,u}^j, \tag{3.31}$$

$$\forall j \in J, \forall u \in V, \forall (u,v) \in E : e_{u,v}^j \leq z_u^j. \tag{3.32}$$

$$\forall j \in J, \forall g_s \in \mathscr{Z}_s : \sum_{(u,v) \in g_s} e_{u,v}^j = 2b_{g_s}^j, \tag{3.33}$$

$$\forall j \in J, \forall g_m \in \mathscr{Z}_m, \forall g_s \in g_m : \sum_{(u,v) \in g_s} e_{u,v}^j \leq 2b_{g_m}^j, \tag{3.34}$$

Constraint (3.31) ensures that a bm-trail traverses a link in both directions. Note that m-trails are formulated as directed paths which cannot traverse at the same anti-parallel links. Equations (3.33), (3.34), (3.32) formulates the same constraints for bm-trails as Eqs. (3.14), (3.15), (3.30) for m-trails, respectively.

3.3 UFL for Multiple Failures

In summary of this section, the linear constraints for single-link, sparse-SRLG and multiple failure localization are presented. Furthermore, monitoring structure formation for both m-trails and bm-trails is introduced.

3.3.4 Sufficient and Necessary Conditions for SRLG UFL

In this section the necessary and sufficient conditions on $\underline{\underline{A}}$ are discussed for code uniqueness (R1) of the MAP problem both for sparse- and dense-SRLGs. Note that for unambiguous single-link failure localization it is sufficient and necessary that the a_e link codes are unique. However, for multiple-link SRLG failure localization, this becomes a nontrivial task due to the requirement that the code of an SRLG is the bitwise OR of the links contained in the SRLG.

The codes of SRLG $z_k = \{k_1, \ldots, k_n\}$ and SRLG $z_l = \{l_1, \ldots, l_m\}$ are different, if and only if there exists a position j in their code that is different, i.e., $a_{\{k_1,\ldots,k_n\},[j]} = 1$ and $a_{\{l_1,\ldots,l_m\},[j]} = 0$ or vice versa. As (b)m-trails are routed over the network links while SRLGs are simply logical entities, we will mainly deal with the conditions on link codes that can support (R1) for the SRLGs.

3.3.4.1 Conditions on Code Uniqueness for Dense-SRLG Model

In this section, the conditions formulated for multiple failures are discussed, when CGT coding techniques are applied. With dense-SRLGs, most of the link sets containing up to arbitrary d links are taken as SRLGs; thus, it is useful to make $\underline{\underline{A}}$ $\bar{d}$-separable to ensure that the bitwise OR of up to arbitrary d codes in $\underline{\underline{A}}$ is unique.

A stronger property than $\bar{d}$-separable is d-disjunct, in which OR of up to d codes does not *contain*[5] any other code. Here we say a code C_1 *contains* code C_2 if for every bit position $c_{1,[i]} \geq c_{2,[i]}$. Lemma 8.1.2 of [13] implies that for a set of $d+1$ d-disjunct codes, any code in the set of codes always has a bit position as 1 while 0 at the same bit position of all the other d codes. The following corollary holds:

Corollary 3.1. *Any arbitrary d rows of a d-disjunct matrix $\underline{\underline{A}}$ contains a d-by-d submatrix, where the submatrix is a permuted d-order identity matrix.*

The corollary is applicable to the dense-SRLG model where all link sets with up to arbitrary d links are considered as SRLGs. In the sparse-SRLG case, nonetheless, a weaker necessary condition can be obtained to satisfy the SRLG code uniqueness requirement, which will be described in the following subsection.

---

[5] In the original terminology, the codes are characteristic vectors of sets.

3.3.4.2 Conditions of Code Uniqueness for Sparse-SRLG Model

Let a set of sparse-SRLGs be defined in the network. Similarly to Lemma 8.1.2 of [13] for sparse-SRLG we have the following lemma:

Lemma 3.5. *A necessary and sufficient condition to distinguish a multi-link SRLG $z_f = \{f_1, f_2, \ldots, f_d\}$, from a single-link failure on $z_i = \{f_i\}$, where $1 \leq i \leq d$, is that there exists a bit position l, $1 \leq l \leq b$, such that the corresponding bit in the link code $a_{f_i,[l]} = 0$, and $\exists k \neq i, 1 \leq k \leq d : a_{f_k,[l]} = 1$.*

Proof. It is clear that $\forall j, 1 \leq j \leq b$, if $a_{f_i,[j]} = 1$ we have $a_{\{f_1,f_2,\ldots,f_d\},[j]} = 1$ due to the bitwise OR operation. Thus, the code for $z_f = \{f_1, f_2, \ldots, f_d\}$ contains the code of $z_i = \{f_i\}$ (i.e., z_f has "1" in the bit position where z_i has "1"). In order to distinguish a_{f_i} and $a_{\{f_1,f_2,\ldots,f_d\}}$, the following must hold: $\exists l, 1 \leq l \leq b$: $a_{\{f_1,f_2,\ldots,f_d\},[l]} = 1$ while $a_{f_i,[l]} = 0$. As $a_{f_1,f_2,\ldots,f_d} = a_{f_1}$ OR a_{f_2} OR $\ldots a_{f_d}$, the only feasible situation is that $\exists k \neq i, 1 \leq k \leq d : a_{f_k,[l]} = 1$.

With Lemma 3.5, we have the following corollary:

Corollary 3.2. *If the alarm codes $a_{f_1}, a_{f_2}, \ldots, a_{f_m}$ of matrix $\underline{\underline{A}}$ contain a permuted m-order identity submatrix, then any single-link SRLG $z_i = \{f_i\}$ has a different alarm code from that of SRLG $z_f = \{f_1, f_2, \ldots, f_m\}$, where $1 \leq i \leq m$.*

Note that the existence of an identity matrix is a sufficient but not necessary condition to the uniqueness of $a_{\{f_1,f_2,\ldots,f_d\}} = a_{f_1}$ OR a_{f_2} OR $\ldots a_{f_d}$ from any $a_{f_k} \forall 1 \leq k \leq d$. A counter example is shown below: with SRLG $z = \{f_1, f_2, f_3\}$ and its $\underline{\underline{A}}$ matrix:

$$\underline{\underline{A}} = \begin{Bmatrix} a_{f_1} \\ a_{f_2} \\ a_{f_3} \end{Bmatrix} = \begin{Bmatrix} 1 & 1 & 0 \\ 0 & 1 & 1 \\ 1 & 0 & 1 \end{Bmatrix}. \tag{3.35}$$

The matrix meets the necessary condition in Lemma 3.5 although it is obviously not an identity matrix. The code uniqueness can be easily verified; for example, since there is a link code with 1 on the third bit, i.e., $a_{f_2,[3]} = 1$, the bitwise OR of a_{f_1}, a_{f_2}, and a_{f_3} must be an alarm code with 1 in its third bit position (i.e., $a_{\{f_1,f_2,f_3\},[3]} = 1$). Further since $a_{f_1,[3]} = 0$, the alarm code of SRLG $z = \{f_1, f_2, f_3\}$ is different from the alarm code of the link f_1.

Also, we can clearly see that Lemma 3.5 is not sufficient for meeting the SRLG code uniqueness requirement because it only ensures the unambiguity between a multi-link and a single-link SRLG contained by the multi-link SRLG, while the unambiguity of two multi-link SRLG codes is not addressed. This will be solved in the next subsection.

3.3.4.3 Strict Sufficient Condition for (R1) for Single- and Multi-Link SRLGs

In this section Theorem 3.9 provides a strict sufficient condition for the SRLG code uniqueness requirement in the sparse-SRLG model, which serves as the foundation of the proposed algorithm presented in Sect. 3.3.5.

Theorem 3.9. *All single-link and considered multi-link SRLGs are uniquely coded if the following three conditions hold for the link codes in $\underline{\underline{A}}$:*

(i) $\forall e \in E : a_e \neq 0\ldots 0$,
(ii) $\forall e, f \in E : a_e \neq a_f$,
(iii) $\forall z \in \mathscr{Z}, z = \{f_1, f_2, \ldots, f_m\}$ and $\forall e \in E$, if $\exists l, \exists i : a_{e,[l]} = a_{f_i,[l]} = 1$, then $\forall k \neq i : \nexists j$ where $a_{e,[j]} = a_{f_k,[j]} = 1$, $1 \leq i, k \leq m$ and $1 \leq l, j \leq b$.

Proof. In a nutshell, the proof is to demonstrate that every SRLG $z_f = \{f_1, f_2, \ldots, f_m\}$ has a unique alarm code provided that the three conditions on the link codes $a_e, e \in E$ are satisfied. We will show that two arbitrary and different SRLGs, denoted as SRLG $z_f = \{f_1, f_2, \ldots, f_m\}$ and $z_e = \{e_1, e_2, \ldots, e_n\}$, respectively, have different codes.

For easy presentation, the SRLGs are represented as a graph $B = (U_B, V_B, E_B)$ shown in Fig. 3.15, where V_B represents the set of links in SRLG $z_f = \{f_1, f_2, \ldots, f_m\}$, U_B represents the set of links in SRLG $z_e = \{e_1, e_2, \ldots, e_n\}$, and E_B shows the interconnection among vertices in B. E_B is defined in such a way that any two vertices w_1 and w_2 in B are connected if $\exists j \in [1, \ldots, b_{max}] : a_{w_1,[j]} = a_{w_2,[j]} = 1$. Thus, the graph B is a bipartite graph, and according to Condition (iii) the codes of two links of a common SRLG cannot be 1 at the same bit position. By Condition (iii), the nodal degree δ of any vertex $u \in U_B$ or $v \in V_B$

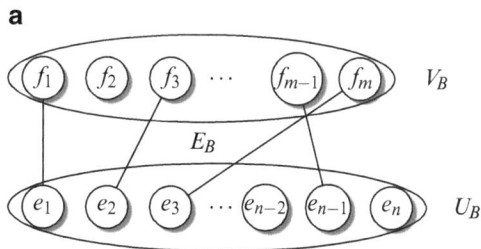

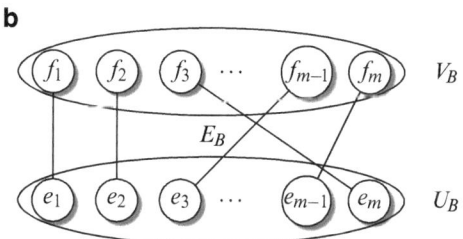

Fig. 3.15 Different scenarios on graph B. (**a**) Some nodes are isolated in B. (**b**) There are no isolated nodes

is 0 or 1. Thus, the edges in E_B define a *matching* between the two vertex sets, U_B and V_B, of the two SRLGs. Two vertices w_1 and w_2 are *matched* if and only if we can find any bit position j such that $a_{w_1,[j]} = a_{w_2,[j]} = 1$.

There are two possible situations according to the matching between the vertices of the two subgraphs in B, which are discussed as follows. The first situation is on partial matching, where at least one vertex in any subgraph has a zero nodal degree. Without loss of generality, we will discuss the case that $\exists e_k \in U_B : \delta(e_k) = 0$. This means that there exists an $e_k \in U_B$ which is not matched by any vertex in V_B. In this case there is an index j, such that $1 \leq j \leq b_{max}$, $a_{e_k,[j]} = 1$ and $a_{f,[j]} = 0$ for every $f \in z_f$. Thus, the jth bit position of SRLG $z_f = \{f_1, f_2, \ldots, f_m\}$ will be 0, while it is 1 for SRLG $z_e = \{e_1, e_2, \ldots, e_n\}$.

The second case is when there is an edge (e_k, f_i) in B. The presence of the edge means that $\exists j \in [1, \ldots, b_{max}] : a_{f_i,[j]} = a_{e_k,[j]} = 1$. From Condition (i) and Condition (ii) we know that each link has a unique nonzero link code, without loss of generality $\exists q \in [1, \ldots, b_{max}] : a_{f_i,[q]} = 1$ and $a_{e_k,[q]} = 0$. From Condition (iii) $\forall l \neq k, 1 \leq l \leq n : a_{e_l,[q]} = 0$. Thus, there will be 1 at the qth bit position of SRLG $z_f = \{f_1, f_2, \ldots, f_m\}$, while 0 for SRLG $z_e = \{e_1, e_2, \ldots, e_n\}$.

We give an example to demonstrate that the condition in the theorem is sufficient but not necessary. The example will show a link code matrix that achieves code uniqueness, but it violates the conditions listed in the theorem. Let us consider all single-link and dual-link SRLGs for the four links with the codes in the following link code matrix:

$$\underline{\underline{\mathbf{A}}} = \begin{Bmatrix} a_{f_1} \\ a_{f_2} \\ a_{f_3} \\ a_{f_4} \end{Bmatrix} = \begin{pmatrix} 1\ 0\ 1\ 0\ 0 \\ 0\ 1\ 0\ 1\ 0 \\ 1\ 0\ 0\ 0\ 1 \\ 0\ 1\ 1\ 0\ 0 \end{pmatrix}. \tag{3.36}$$

One can easily check that the code uniqueness holds for all the single-link and dual-link SRLGs. On the other hand, e.g., SRLG $z = \{f_1, f_2\}$ does not satisfy Condition (iii) because in the link code matrix in Eq. (3.36) exists a link code a_{f_4}, which has two positions, namely $l = 2$ and $j = 3$, where a_{f_1} and a_{f_2} from the SRLG $z = \{f_1, f_2\}$ both have bit "1."

It is possible to find a less strict sufficient condition than that by Theorem 3.9; however, the one in Theorem 3.9 is simple and allows a fast and efficient implementation of m-trail code assignment for $\underline{\underline{A}}$ with alarm code uniqueness of each SRLG. By using the condition in Theorem 3.9 for general multi-link SRLGs, we develop an algorithm for m-trail allocation in presence of SRLGs containing a single-link or multiple adjacent links.

3.3 UFL for Multiple Failures

Fig. 3.16 Different rules on the link codes in the graph-representation of two SRLGs. (**a**) Strict rule on link codes in Theorem 3.9. (**b**) Permissive rule on link codes in Theorem 3.10

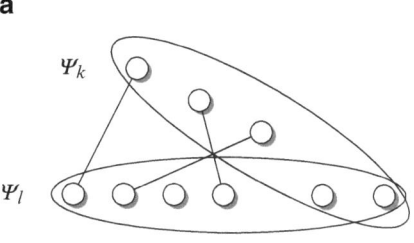

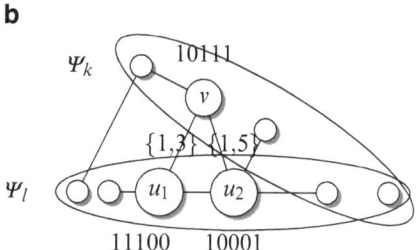

3.3.4.4 Permissive Necessary and Sufficient Condition for (R1)

In the following, a similar auxiliary graph to represent the SRLGs is introduced as in Fig. 3.15, which will be used in the problem formulation of a more general permissive condition on the link codes. Let Ψ_i be a node set, in which each node represents a link in SRLG z_i. For example, if SRLG z_k consists of two edges $z_k = \{e, f\}$, then Ψ_k contains two nodes e and f. In a graph $S = (\Psi_k \cup \Psi_l, E_S)$ two nodes v and u (corresponding to links e and f in the original graph) are connected, if and only if $\exists j : a_{e,[j]} = a_{f,[j]} = 1$. Each link is labeled with the positions $\{j_1, \ldots, j_m\}$, where $a_{e,[j_i]} = a_{f,[j_i]} = 1$. Let $c(v)$ denote the characteristic vector of a link code a_v, which characterizes the bit positions where a_v is "1." For example, if $a_v = 10111$, then $c(v) = \{1, 3, 4, 5\}$. Furthermore, $c(v, u)$ is defined as $c(v, u) = c(v) \cap c(u)$, and $N(v)$ is defined for node $v \in \Psi_i$ as $N(v) = \cup_{\forall u \in \Psi_j} c(v, u)$. For example, for node v the set is $N(v) = \{1, 3, 5\}$ in Fig. 3.16b. Obviously, $N(v) \subseteq c(v)$ for all v.

The necessary and sufficient conditions are formulated accordingly as follows:

Theorem 3.10. *The codes of arbitrarily chosen two SRLGs ($z_k = \{k_1, \ldots, k_n\}$ and $z_l = \{l_1, \ldots, l_m\}$) are different, if and only if $\exists v \in \{(\Psi_k \cup \Psi_l) \setminus (\Psi_k \cap \Psi_l)\} : N(v) \subset c(v)$.*

Proof. Without loss of generality, $v \in \Psi_k \setminus (\Psi_k \cap \Psi_l)$. The condition is sufficient, as $a_{\{k_1,\ldots,k_n\},[j]} = 1$ for all $j \in c(v) \setminus N(v)$, while $a_{\{l_1,\ldots,l_m\},[j]} = 0$.

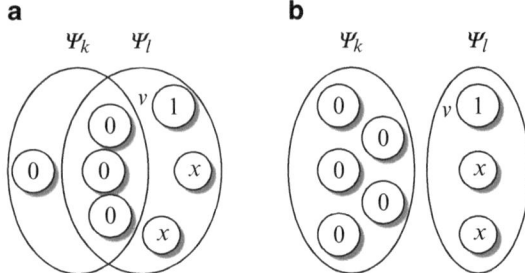

Fig. 3.17 Satisfaction of the strong unambiguity rule on link v and position j. The values $a_{\{\Psi_k\},[j]} = 0$, $a_{\{\Psi_l\},[j]} = 1$ remain the same, regardless of the future assignment of the don't care bits. (**a**) Intersecting SRLGs, (**b**) disjoint SRLGs

The necessity part is shown below. Let us assume that $a_{\{k_1,\ldots,k_n\}} = a_{\{l_1,\ldots,l_m\}}$. By possibly swapping the indices k and l we may assume that there is a j such that $a_{\{k_1,\ldots,k_n\},[j]} = 1$ and $a_{\{l_1,\ldots,l_m\},[j]} = 0$. Then there is an i such that $a_{k_i,[j]} = 1$, while for $1 \leq s \leq n$ we have $a_{l_s,[j]} = 0$. This implies that $k_i \in \Psi_k \setminus \Psi_l$, and $j \in c(k_i)$ but $j \notin N(k_i)$, hence $N(k_i) \subset c(k_i)$. The proof is complete.

We give an example to demonstrate the theorem. Let node $v \in \Psi_k$ have a link code $a_v = 10111$ while $c(v, u_1) = \{1, 3\}$ and $c(v, u_2) = \{1, 5\}$ in Fig. 3.16b. In the example, $N(v) = \{1, 3, 5\} \subset c(v) = \{1, 3, 4, 5\}$. Thus, the two SRLG codes are different on the 4th bit position, i.e., $a_{\{k_1,\ldots,k_n\},[4]} = 1$ while $a_{\{l_1,\ldots,l_m\},[4]} = 0$.

Corollary 3.3 (Strong Unambiguity Rule). *If there exists a link code a_v of an arbitrary link v from the symmetric difference ($v \in \{(\Psi_k \cup \Psi_l) \setminus (\Psi_k \cap \Psi_l)\}$) of SRLG Ψ_k and SRLG Ψ_l, without loss of generality for $v \in \Psi_l : a_{v,[j]} = 1$, and $\forall u \in \Psi_k : a_{u,[j]} = 0$, then the alarm codes of the SRLGs are different.*

If the strong unambiguity rule is satisfied on link v and position j, it is easy to check that $j \in c(v)$ and $j \notin N(v)$, thus, $N(v) \subset c(v)$, that satisfies Theorem 3.10. In Fig. 3.17 two examples are shown for the possible situations using the previously introduced graph representation of the SRLGs.

Note that Theorem 3.10 results in similar structures as provided in Theorem 1 in [1]. However, Theorem 1 in [1] cannot help the design of a corresponding scheme for code construction. Thus, the main difference between the two theorems lies in their perspective toward the targeted scenario: Theorem 1 in [1] formulates a rule on the monitoring structures, while Theorem 3.10 provides a rule on the link codes. We will present the proposed Link Code Construction (LCC) heuristic in Sect. 3.3.6 according to the proposed theorem. In a nutshell, in LCC at the beginning the link code matrix $\underline{\underline{A}}$ contains codes with arbitrary nonzero Hamming distances without considering any multi-link SRLG, thus, code uniqueness in $\underline{\underline{A}}$ is not necessarily ensured. The proposed LCC iteratively extends $\underline{\underline{A}}$ with additional columns j in order to distinguish each SRLG pair in (R1) by using Corollary 3.3. Each element in the added column j in the link code matrix (i.e., $\underline{\underline{A}}_{i,j}$) could be either "1," "0," or don't care ("x"). After the code uniqueness is ensured, bm-trails are formed in each bit position in (R2).

Further note that LCC leaves don't care bits in the jth column after two SRLGs are distinguished using the strong unambiguity rule, only the minimal number of x bits are set. On the other hand, the algorithm following the first design category using Theorem 1 in [1] always sets all bits in the jth bit position, leaving no further space to improve the performance, which is present in LCC.

3.3.5 The Adjacent Link Failure Localization Heuristic Approach

In this section, the MAP problem for UFL of adjacent link failures using m-trails and bm-trails is investigated. In the single-link failure scenario, an ILP can be formulated and used to solve small-sized problems with reasonable computation time [44]. In the multiple failure case, the solving of an ILP formulation is intractable even for small inputs of the problem because of the additional bitwise OR operations on the link codes (see Sect. 3.3.3). Thus, an intelligent algorithm is proposed, called *Adjacent Link Failure Localization (AFL)*, to achieve a fast yet efficient m-trail solution for UFL of SRLGs containing a single-link or multiple adjacent links. Since the failure of all adjacent links implicitly means the failure of their common node, the proposed m-trail solution can be generally claimed for *node failure monitoring*. Due to the possibly highly heterogeneous SRLGs in terms of the number of links, the approaches designed for the dense-SRLG scenario, such as CGT or as that in [18], would be very inefficient.

In this section, the problem of m-trail allocation is formulated into three sub-tasks that will be performed one after the other. The first is to *partition the graph into subgraphs* in an *(R0)* step, where the m-trail formation is performed independently. In the second stage, for each partition *code assignment (R1)* is performed, where each link is first (or tentatively) assigned with a unique alarm code under some given constraints. The third stage is the *m-trail formation (R2)*, which ensures that we can find a set of non-simple paths as m-trails, such that the jth m-trail traverses through all the links with the bit at the jth bit position as "1" while disjoint from any link with the bit at the jth bit position as "0."

The basic idea of the proposed AFL algorithm is to divide the whole MAP into sub-problems, such that each sub-problem can be solved as a single-link UFL case. The division of the problem is by way of a topology partitioning method that partitions the whole topology into subgraphs according to the set of SRLGs, such that the dependency of link codes in each partition is removed. Thus, the m-trail formation can be performed via RCS (see Sect. 3.2.7) in each partition, or bm-trail allocation via greedy code swapping (GCS) (see Sect. 3.3.7). The complete m-trail solution is given by taking all the m-trails in the subgraphs (as the direct sum of the $\underline{\underline{A}}_i$ matrices obtained in the partitions).

3.3.5.1 Topology Partitioning for SRLG Code Uniqueness

The topology partitioning method, which aims to meet the sufficient condition of SRLG code uniqueness defined in Theorem 3.9, is presented in this subsection. Let the network topology be denoted as $G = (V, E)$. The pseudo code of the proposed method is presented in Algorithm in 1. In Step (3), the *line graph* of G is formed, denoted as $L(G) = (V_{L(G)}, E_{L(G)})$, which is constructed in such a way that each vertex of $L(G)$ represents an edge of G. Let the set of vertices and edges of $L(G)$ be denoted by $V_{L(G)}$ and $E_{L(G)}$, respectively. Thus, any two vertices in $V_{L(G)}$ are adjacent if and only if their corresponding links in G are incident to a common node. Figure 3.18a, b show an example, where the graph G with 4 nodes and 5 links is transferred to a line graph $L(G)$ with 5 vertices and 8 edges.

The topology partitioning process runs on $L(G)$. Each edge in $L(G)$ is marked by red (or blue) if the two end vertices of the edge, which represent two links in G, belong (or do not belong) to a common SRLG (as indicated in Step (6) and Step (10)). An example is shown in Fig. 3.18a where all the single-link SRLGs and all adjacent dual-link SRLGs, except SRLG $z_1 = \{(0, 1), (1, 3)\}$ and

---

Algorithm 1: Adjacent-link Failure Localization (AFL) Algorithm

Input: $G = (V, E), \mathscr{Z}$
Result: $\underline{\underline{A}}$

1 **begin**
2   Initialize $\underline{\underline{A}}$ empty;
3   Form the line graph $L(G) = (V_{L(G)}, E_{L(G)})$;
4   **for** $v = (e, f) \in E_{L(G)}$ **do**
5   **if** $\exists z_i \in \mathscr{Z} : e, f \in z_i$ **then**
6   color $v = (e, f) \in E_{L(G)}$ red;
7   **end**
8
9   **else**
10   color $v = (e, f) \in E_{L(G)}$ blue;
11   **end**
12
13   **end**
14   **while** $\exists v \in E_{L(G)}$ *with blue color* **do**
15   Take a maximum vertex-induced subgraph $L(H)$ with blue links;
16   Use RCS (or GCS) to form m-trails (bm-trails) in H, results $\underline{\underline{A}}_H$;
17      $\underline{\underline{A}} := \underline{\underline{A}} \oplus \underline{\underline{A}}_H$;
18      $\forall v = (e, f) \in E_{L(H)}$: delete v from $E_{L(G)}$;
19   **end**
20   **while** $\exists e \in E : a_e = [0, \ldots, 0]$ **do**
21      $\underline{A}_e = [1]$;
22      $\underline{\underline{A}} := \underline{\underline{A}} \oplus \underline{A}_e$;
23   **end**
24 **end**

3.3 UFL for Multiple Failures

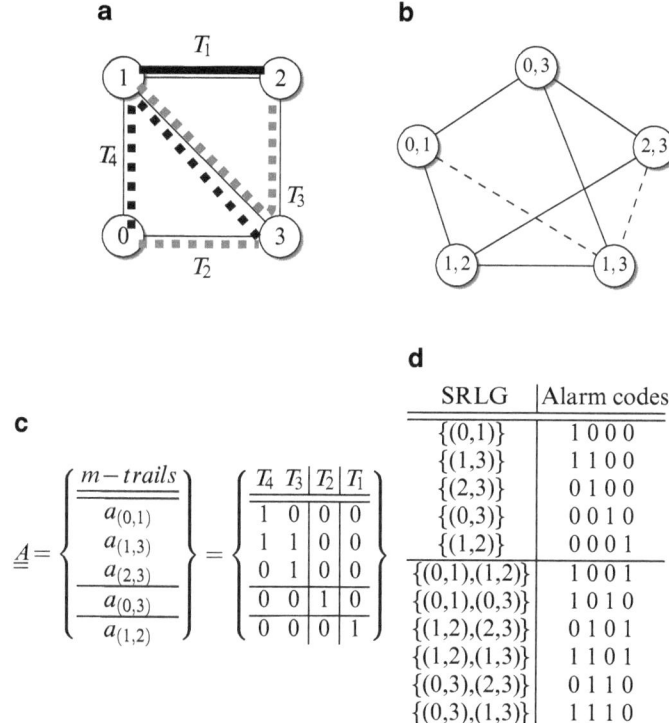

Fig. 3.18 An example on link code assignment and resulting alarm code table with the AFL algorithm. (**a**) Four-node mesh network G and the m-trail solution, (**b**) line graph $L(G)$ of G, (**c**) link code matrix, (**d**) SRLG codes

$z_2 = \{(2, 3), (1, 3)\}$, are considered. The line graph $L(G)$ is given in Fig. 3.18b. The edge $v_1 = ((0, 1), (0, 3)) \in E_{L(G)}$ is an edge in $L(G)$, and $(0, 1)$ and $(0, 3)$ belong to $z_1 = \{(0, 1), (0, 3)\}$. Thus, v_1 is colored by red (solid line in the figure) to indicate that $(0, 1)$ and $(0, 3)$ should not be in the same partition so as to maintain the independence of the codes at each partition. On the other hand, the edge $v_2 = ((0, 1), (1, 3)) \in E_{L(G)}$ is colored by blue (dashed line in the figure) since the two links in G represented by the vertices $(0, 1)$ and $(1, 3)$ do not belong to a common SRLG.

In the loop of Steps (15)–(18), the algorithm iteratively identifies each maximal vertex-induced subgraph where all edges colored with blue in $L(G)$, and restores it back to the original graph domain (denoted as H in Algorithm 1). Then, RCS (or GCS) is performed on H. The m-trail formation on H results in an alarm code matrix $\underline{\underline{A}}_H$, which will be *direct summed* with the existing alarm code matrix $\underline{\underline{A}}$. The direct sum $\oplus$ of arbitrary matrices (not necessary quadratic and the same size) $\underline{\underline{A}}_1, \underline{\underline{A}}_2, \ldots, \underline{\underline{A}}_n$ is a block matrix formed as

$$\underline{\underline{A}} = \bigoplus_{i=1}^{n} \underline{\underline{A}}_i = \left\langle \underline{\underline{A}}_1, \underline{\underline{A}}_2, \ldots, \underline{\underline{A}}_n \right\rangle = \begin{Bmatrix} \underline{\underline{A}}_1 & 0 & 0 & 0 \\ 0 & \underline{\underline{A}}_2 & 0 & 0 \\ 0 & 0 & \ddots & 0 \\ 0 & 0 & 0 & \underline{\underline{A}}_n \end{Bmatrix},$$

where 0 denotes a matrix of proper size whose elements are all "0."

As a result, the final alarm code matrix $\underline{\underline{A}}$ will contain blocks along its diagonal, by which the requirement of Corollary 3.2 can always be fulfilled. At the end of the loop, the edges of $L(H)$ are removed from $L(G)$ as shown in Step (18).

The above process iterates on all the blue maximum vertex-induced subgraphs one after the other. Finally, the edges $e \in E$ in G, which were not involved in any H subgraph during the iterations in Steps (15)–(18), will be taken one by one as a partition and assigned by [1] as the 1×1 alarm code matrix $\underline{\underline{A}}_e$ in Step (21). Figure 3.18c gives an example on the block diagonal matrix while unique SRLG codes contained in the corresponding $\underline{\underline{A}}$ is shown in Fig. 3.18d.

The following theorem proves that the proposed partitioning method can remove the code dependency in each partition. This implies that in any subgraph of G corresponding to a partition only single-link SRLGs need to be handled. After coding the partitions independently with methods proposed for single-link SRLG case the final m-trail solution can be obtained by simply including all the m-trails generated in each subgraph, which is equivalent with the direct sum of the partitions link code matrices.

Theorem 3.11. *The partitioning and coding method proposed in Algorithm 1 can sufficiently achieve SRLG code uniqueness.*

Proof. In a nutshell, the proof is to demonstrate that the resulting $\underline{\underline{A}}$ link code matrix meets all the conditions of Theorem 3.9.

According to the edge coloring in the proposed partitioning process, a red edge will be between two vertices of $L(G)$ which represents arbitrary two links f_i and f_j belonging to SRLG $z_f = \{f_1, f_2, \ldots, f_m\}$. Therefore, the two links must be coded in different partitions. Thus, an arbitrary link e is in the same partition with at most one link from an arbitrary SRLG $z_f = \{f_1, f_2, \ldots, f_m\}$, without loss of generality f_s (or with itself, if e is part of the SRLG). As the final link code matrix $\underline{\underline{A}}$ is the direct sum of the partition codes $\underline{\underline{A}}_k$, e will have 0 in all bit positions, where all other f_i links $i \neq s$ could have 1. Thus, e can have 1 in the same bit position at most f_s (or itself) from the SRLG $z_f = \{f_1, f_2, \ldots, f_m\}$, which satisfies Condition (iii) of Theorem 3.9.

Note that we can trivially keep $a_e \neq 00\ldots0$ and $a_e \neq a_f$ in the partitions with single-link SRLG methods (like RCS), which satisfy Conditions (i) and (ii) in the subgraphs and after direct summing, in $\underline{\underline{A}}$ too. Therefore, with the proposed method all the three conditions defined in Theorem 3.9 can be satisfied, and the code uniqueness of all the SRLGs can be achieved.

3.3 UFL for Multiple Failures

There are two borderline cases of the proposed partitioning method. First, in the event that only single-link SRLGs are considered, the whole line graph is blue and the problem can be solved directly by RCS. On the other hand, when there is no more than one blue edge in any $L(H)$, the resultant alarm code matrix will be an $|E|$-order identity matrix, which means the solution simply leads to link-based monitoring. In this sense, the proposed algorithm can outperform link-based monitoring (i.e., taking less than $|E|$ monitors) if there exists at least one maximal vertex-induced blue subgraph with two or more blue edges, as shown in the example in Fig. 3.18.

An AFL solution of Algorithm 1 is presented in Fig. 3.19 for SmallNet for sparse-SRLG failures. Based on the sparse-SRLG list to be localized (shown in Fig. 3.19), the partitioning method runs in three iterations. First, 16 links are coded together with GCS (resulting bm-trails $T_1 - T_5$). In the next two iterations 3–3 links are coded together, both forming a maximal vertex-induced blue subgraph with two blue edges, resulting bm-trails $T_6 - T_7$ and $T_8 - T_9$, respectively. Although built on the strict conditions formulated in Theorem 3.9, it is worth to mention that both the number of bm-trails (from 12 to 9) and the cover length of the solution (from 81 to 63) are decreased with AFL compared to the CGT approach in Fig. 3.14. Thus, we can conclude that although methods like CGT can efficiently handle multiple failure scenarios, these are not necessarily efficient to localize sparse-SRLGs. On the other hand, for the sparse-SRLG problem, algorithms built on the characteristics of this special case—like AFL—can provide much better solutions.

3.3.5.2 Complexity Analysis

Creating the line graph in Step (3) takes $O(|E| \cdot \Delta)$ complexity, while coloring the edges in Steps (6) and (10) takes $O(|E|\Delta \cdot |SRLG|\Delta)$, where Δ denotes the maximal nodal degree of graph G, and $|SRLG|$ is at most $O(|E|\Delta + |V|)$ in sparse-SRLG model. The worst complexity of the implemented heuristic for finding the vertex-induced subgraphs in Step (15) is $O(|E|\Delta \cdot |E|)$ since we take $v \in E_{L(G)}$ and do a greedy check on all the other $e \in V_{L(G)}$ no matter it has a red edge or not. Finally, the worst case complexity of using RCS in Step (16) is $O(|V| \cdot |E|^2 \cdot \Delta^6 \cdot \log^3 |E|)$, and it is used at most $|E|$ times. Thus, the code assignment complexity in Steps (15)–(18) is $O(|V| \cdot |E|^3 \cdot \Delta^6 \cdot \log^3 |E|)$, while the code assignment for each individual link in Step (21) at the end is $O(|E|)$. Based on the above, the overall complexity is the complexity of the iteration Step (15)–(18) with $O(|V| \cdot |E|^3 \cdot \Delta^6 \cdot \log^3 |E|)$.

3.3.6 The LCC Heuristic Approach

As the bm-trail allocation for SRLGs has been shown to be NP-complete, in the following a heuristic approach is proposed for UFL of arbitrary SRLG lists $\mathscr{L}$ (including multiple failures) following the permissive necessary and sufficient condition in Theorem 3.10.

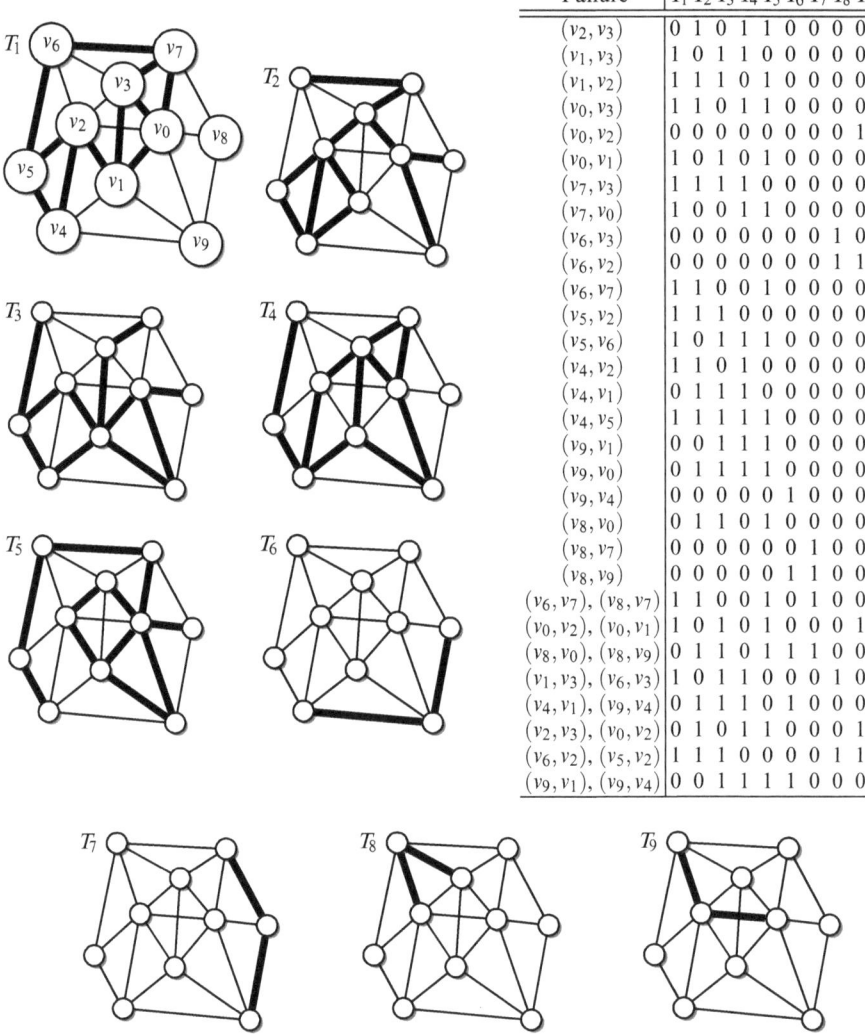

Failure	T_1 T_2 T_3 T_4 T_5 T_6 T_7 T_8 T_9
(v_2, v_3)	0 1 0 1 1 0 0 0 0
(v_1, v_3)	1 0 1 1 0 0 0 0 0
(v_1, v_2)	1 1 1 0 1 0 0 0 0
(v_0, v_3)	1 1 0 1 1 0 0 0 0
(v_0, v_2)	0 0 0 0 0 0 0 0 1
(v_0, v_1)	1 0 1 0 1 0 0 0 0
(v_7, v_3)	1 1 1 1 0 0 0 0 0
(v_7, v_0)	1 0 0 1 1 0 0 0 0
(v_6, v_3)	0 0 0 0 0 0 0 1 0
(v_6, v_2)	0 0 0 0 0 0 0 1 1
(v_6, v_7)	1 1 0 0 1 0 0 0 0
(v_5, v_2)	1 1 1 0 0 0 0 0 0
(v_5, v_6)	1 0 1 1 1 0 0 0 0
(v_4, v_2)	1 1 0 1 0 0 0 0 0
(v_4, v_1)	0 1 1 1 0 0 0 0 0
(v_4, v_5)	1 1 1 1 1 0 0 0 0
(v_9, v_1)	0 0 1 1 1 0 0 0 0
(v_9, v_0)	0 1 1 1 1 0 0 0 0
(v_9, v_4)	0 0 0 0 0 1 0 0 0
(v_8, v_0)	0 1 1 0 1 0 0 0 0
(v_8, v_7)	0 0 0 0 0 0 1 0 0
(v_8, v_9)	0 0 0 0 0 1 1 0 0
$(v_6, v_7), (v_8, v_7)$	1 1 0 0 1 0 1 0 0
$(v_0, v_2), (v_0, v_1)$	1 0 1 0 1 0 0 0 1
$(v_8, v_0), (v_8, v_9)$	0 1 1 0 1 1 1 0 0
$(v_1, v_3), (v_6, v_3)$	1 0 1 1 0 0 0 1 0
$(v_4, v_1), (v_9, v_4)$	0 1 1 1 0 1 0 0 0
$(v_2, v_3), (v_0, v_2)$	0 1 0 1 1 0 0 0 1
$(v_6, v_2), (v_5, v_2)$	1 1 1 0 0 0 0 1 1
$(v_9, v_1), (v_9, v_4)$	0 0 1 1 1 1 0 0 0

Fig. 3.19 An AFL heuristic solution based on bm-trails for sparse-SRLGs. The solution has $b = 9$ and $\|\underline{\mathscr{T}}\| = 63$

In LCC each SRLG pair is considered sequentially in $\mathscr{L}$, where their codes are derived using the bitwise OR operation on the codes of the contained links. Then the algorithm checks whether their codes collide or not. If yes, it is resolved by extending the $\underline{A}$ matrix with a column (e.g., column j) with don't care bits, and apply a minimal cost code assignment following the strong unambiguity rule in Corollary 3.3. To be specific, with any SRLG code collision of SRLGs $z_k = \{k_1, k_2, \ldots, k_n\}$ and $z_l = \{l_1, l_2, \ldots, l_m\}$, the two SRLG alarm codes are made

3.3 UFL for Multiple Failures

unique in $\underline{\underline{A}}$ by ensuring the existence of a position j, where $a_{\{k_1,k_2,...k_n\}.[j]} = 1, a_{\{l_1,l_2,...l_m\}.[j]} = 0$ or vice versa. In order to do so, we have to keep in mind the bitwise OR relation between the link codes and SRLGs codes. Hence, if any don't care bit in the position j of alarm code of a link contained by SRLG z_k is set in the future (either "1" or "0"), e.g., $a_{k_d.[j]} = x$ is set to 1, the jth position in the alarm code of SRLG z_k remains the same $a_{\{k_1,k_2,...k_n\}.[j]} = 1$. However, if a link code contained by SRLG z_l is set to 1 in the further iterations, e.g., $a_{l_d.[j]} = x$ is set to 1, it changes the jth position of SRLG alarm code z_l to 1, i.e., $a_{\{l_1,l_2,...l_m\}.[j]} = 1$. Such change can possibly make the codes of the two SRLGs identical. If the strong unambiguity rule formulated in Corollary 3.3 is true for the two SRLGs, then the alarm codes of the two SRLGs remain different regardless of the future assignment of the don't care bits.

The pseudo code of the proposed LCC algorithm is presented in Algorithm 2. In Step (2) single-link UFL is ensured with RCS. In Iterations (5) and (6), the algorithm compares every pair of SRLG alarm codes, and in case of any code collision, the algorithm makes them different from each other using the strong unambiguity rule shown in Fig. 3.17. Note that each SRLG pair is checked only once, and their codes remain different regardless of the subsequent iterations. To determine which link e, SRLG Ψ_i and bit position pos to perform the strong unambiguity rule, the following SetCost function is used:

$$\text{SetCost}(e, \Psi_i, pos) = \#S + \delta \times \#O,$$

where $\#S$ refers to the number of sets of don't care bits to 1 or 0, and $\#O$ refers to the case when we apply the strong unambiguity rule on a newly added don't care column j in $\underline{\underline{A}}$ which has only don't care bits. The scaling factor δ weights the relative importance of adding a new column in the link code matrix. Since the goal of the method is to minimize the number of bm-trails, that is strongly related to the number of columns used for UFL in $\underline{\underline{A}}$, thus we simply take δ large enough to avoid unnecessary addition of new columns.

In the initial code assignment phase, bm-trail formation was performed in Step (2) such that each column in the initial $\underline{\underline{A}}$ matrix corresponds to an eligible bm-trail. However, further efforts on the newly added columns in Step (8) are needed to minimize the number of bm-trails. Thus, Iteration (12) sets the don't care bits to 1 in the newly added columns in Step (8) in order to create larger connected components on each bit position. Columns containing only don't cares are removed from $\underline{\underline{A}}$.

Finally, in Iteration (21) in a post-process phase (R2) is conducted, i.e., if in a newly added column j in Step (8) cannot lead to a single trail, new columns $a^{b_{pp}+1}$ are added to $\underline{\underline{A}}$. Note that if SRLG Ψ_i and SRLG Ψ_j are distinguished on the jth position in Step (8), after (R2) in Iteration (21), the strong unambiguity remains true either on the jth position or on the newly added $a^{b_{pp}+1}$ column. As a result, all SRLGs in the SRLG list $\mathscr{L}$ are unambiguously localized, i.e., each code in the alarm code table derived from the link code matrix $\underline{\underline{A}}$ constructed with Algorithm 2 is unique.

Algorithm 2: Link Code Construction (LCC) Algorithm

Input: $G = (V, E)$, $\mathscr{L}$
Result: $\underline{\underline{A}}$

1 **begin**
2   Use RCS to form bm-trails in G, results $\underline{\underline{A}}_s$ with b_s bit positions ensuring single failure localization;
3     $\underline{\underline{A}} := \underline{\underline{A}}_s$;
4   Extend $\underline{\underline{A}}$ with $b_{max} - b_s$ don't care columns;
5   **for** $z_k \in \mathscr{L}$ **do**
6   **for** $z_l \in \mathscr{L}$ **do**
7   **if** $k < l$ and $a_{\{k_1,\ldots,k_n\}} = a_{\{l_1,\ldots,l_m\}}$ **then**
8   Set x bits in $\underline{\underline{A}}$ on the bit position pos to $0 \ \forall f \in z_i$ and to 1 for $e \in z_j, i, j \in \{l, k\}$, where:
$$\min{}_{\forall e \in \Psi_j \setminus \Psi_i,\ \forall pos \in J}\ \texttt{SetCost}(e, \Psi_i, pos);$$
9   **end**
10   **end**
11   **end**
12   **for** $j = b_s + 1, \ldots, b_{max}$ **do**
13   **if** $\exists i = 1, \ldots, |E| : \underline{\underline{A}}_{i,j} \neq x$ **then**
14   Set all x on the jth position to 1;
15   **end**
16   **else**
17   Remove column j from $\underline{\underline{A}}$;
18   **end**
19   **end**
20     $b_{ca} :=$ number of columns in $\underline{\underline{A}}$ after code assignment;
21   **for** $j = b_s + 1, \ldots, b_{ca}$ **do**
22        $b_{pp} := b_{ca}$;
23        $CC :=$ The number of connected components in the jth position on the 1 bits;
24   **if** $CC > 1$ **then**
25   **for** $c = 2, \ldots, CC$ **do**
26   Add a column $b_{pp} + 1$ to $\underline{\underline{A}}$, where the elements are set $\forall e_{i_1}, \ldots, e_{i_s} \in c : \underline{\underline{A}}_{i_d,\{b_{pp}+1\}} = 1$, and 0 otherwise;
27   Set $\forall e_{i_1}, \ldots, e_{i_s} \in c : \underline{\underline{A}}_{i_d,j} = 0$;
28                 $b_{pp} := b_{pp} + 1$;
29   **end**
30   **end**
31   **end**
32 **end**

3.3 UFL for Multiple Failures

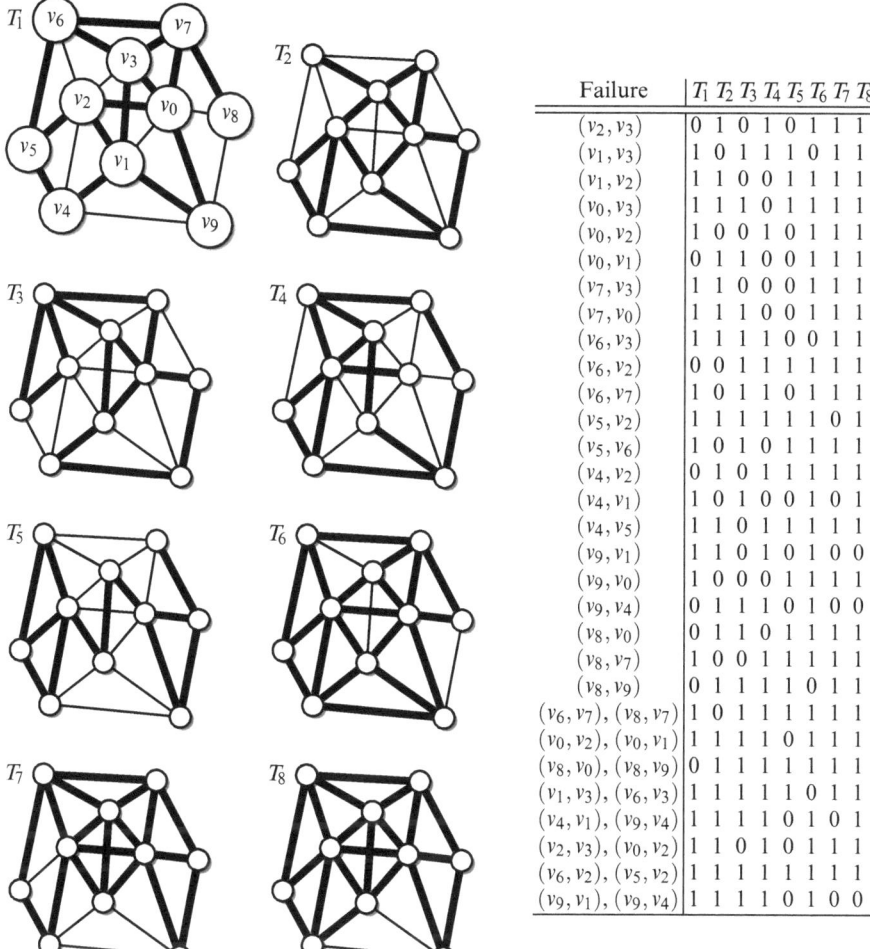

Fig. 3.20 An LCC heuristic solution based on bm-trails for sparse-SRLGs. The solution has $b = 8$ and $\|\mathcal{T}\| = 124$

An example of the LCC heuristic approach is presented in Fig. 3.20 for the same problem instance as for AFL in Fig. 3.19. The initial solution ensuring single-link UFL with GCS has five bm-trails ($T_1 - T_5$). One can observe that the total number of bm-trails required for sparse-SRLG UFL with LCC is less than in the AFL approach ($b = 9$), owing to the difference between the strict rule and permissive rule formulated on the link codes in Fig. 3.16. It is also important to note that as the don't care bits are set to "1" in Step (14) in order to form larger connected components—i.e., less bm-trails—thus, the cover length is almost twice as much as of the AFL approach. Furthermore, there are quite lengthy bm-trails in LCC on these bit positions ($T_6 - T_8$). However, the goal of LCC is only to demonstrate the effectiveness of the permissive rule formulated in Theorem 3.10 to minimize the number of bm-trails. Note that any other rule satisfying the conditions could be applied when the don't care bits are set, which could be based on several operational premises reducing bm-trail length on these bit positions.

3.3.6.1 Receiver and Transmitter Placement

A node failure is equivalent to the failure of all the adjacent links. However, the localization of a node failure concerns more issues than simply localizing a failure event which hits all the adjacent links.

As the LCC design focuses on bm-trails, which are simply a directed Eulerian cycles, and the transmitter and receiver can be possibly placed at the same monitoring location. If a monitoring location fails, it not only makes all the adjacent links unconnected but also fails to perform expected alarm dissemination. Therefore, a backup monitoring location (BML) should be in place along the bm-trail that can identify the status of ML. As both the monitoring location and BML can localize exactly the same failures (they are monitoring the status of the same bm-trail), they issue an alarm at the same time. The alarm of ML suppresses the alarm of BML, as in this case the primary monitoring location is functional. However, if the BML senses a failure, but has not received the alarm of ML, then BML issues the alarm instead of ML.

As a bm-trail traverses at least two nodes, the selection of the primary and BMLs is always possible, allowing full coverage of node failure localization.

3.3.7 The CGT-GCS Heuristic Approach for M-Trail Allocation

The proposed bm-trail allocation method for multiple failures follows the second design principle, where CGT codes generated by [14, 19] are assigned to each link to ensure (R1). After the RCA, (R2) is pursued by way of GCS. Although seemingly similar to that in Sect. 3.2.7, the proposed approach is much different in both stages of code generation and code swapping [36].

Figure 3.21 is a flowchart that summarizes the proposed approach. In Step (1), the CGT code construction GEN_CGT generates a number of k $\overline{d}$-separable codes of a length of J bits, denoted as $\underline{\underline{C}}$. The input parameter $|E|$ ensures that the code length J is the smallest such that $k \geq |E|$. Note that the property of $\overline{d}$-separability ensures uniqueness of the bitwise OR of up to d codes in set $\underline{\underline{C}}$, which is required in (R1). In Step (2), an alarm code table is formed by randomly selecting $|E|$ out of

3.3 UFL for Multiple Failures

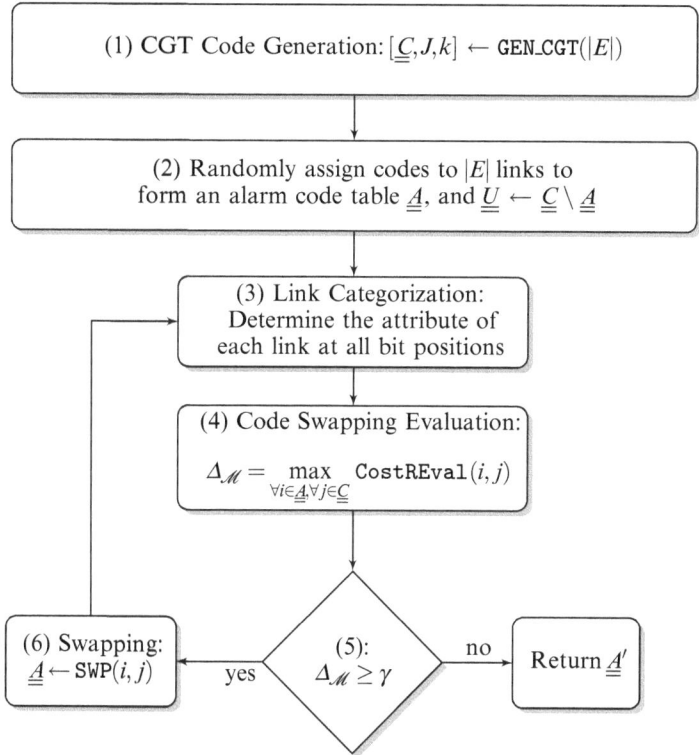

Fig. 3.21 The flowchart for the proposed *CGT-GCS* heuristic algorithm

k codes from $\underline{C}$, which are further assigned to all the links. The group of $|E|$ codes taken by the links is denoted as $\underline{A}$, while the group of remaining $k - |E|$ unassigned codes is denoted as code set $\underline{U}$, where $\underline{U} = \underline{C} \setminus \underline{A}$.

With a CGT code of length J at each link, the best situation is that each bit position of the link can lead to an m-trail, and in this case there are totally J m-trails corresponding to the code assignment. But this is not likely to happen due to the random assignment of the codes at the beginning. Our method solves the m-trail formation problem by GCS starting in Step (3), which ensures (R2).

In Step (3), each link is categorized with one of the four attributes (i.e., isolated, leaf, bridge, and detour) in each bit position according to $\underline{A}$. Next in Step (4), code pair (a_e, C_x), where $a_e \in \underline{A}$ and $C_x \in \underline{C}$, is arbitrarily selected and checked in function CostREval one by one to see how much cost reduction can be achieved by possibly swapping each code pair. The code pair with the steepest cost reduction after swapping is kept (i.e., which minimizes $\Delta_{\mathcal{M}}$). If $\Delta_{\mathcal{M}}$ is no less than γ and at least one bm-trail can be merged or removed, the two codes are swapped using function SWP, such that $\underline{A}$ is updated accordingly in Step (6) and the program then goes back to Step (3). Otherwise, the program returns the best result (i.e., $\underline{A}$ with the least number of bm-trails) and terminates.

Note that an eligible code swapping could be either a swapping between the codes both in $\underline{\underline{A}}$ (i.e., $C_x \in \underline{\underline{A}}$) or replacement of the link code with an unused one (i.e., $C_x \in \underline{\underline{U}}$). Steps (3), (4), (5), and (6) form a loop such that the largest cost reduction can be achieved in each iteration of code swapping.

3.3.7.1 Greedy Code Swapping

To carry out m-trail formation, GCS is devised to greedily swap codes of two links such that the coverage constraint at each bit position can be satisfied while the resultant solution quality can be progressively improved according to the cost function in Eq. (3.1). Such an iterative swapping process continues until a given condition is satisfied.

The cost reduction evaluation for each code swapping serves as an important building block in the proposed GCS mechanism, which guides the m-trail formation process at each link set. In swapping each code pair of two links, a set of regulations is necessary, and will be detailed in the following paragraphs.

The flowchart of the proposed GCS is given in Fig. 3.22, which provides all details of Step (4) in Fig. 3.21. At the beginning, the program picks up a code pair a_e and C_x as shown in Step (4.1), where a_e is a code assigned to link e while C_x is randomly selected from $\underline{\underline{C}}$, respectively. The cost reduction evaluation for a single swapping should be iterated on each bit position (or, each link set) affected by the swapping. The ith bit position (or link set), L_i, is not affected by the swapping of a_e and C_x if the two codes have a common ith bit, i.e., $a_{e,[i]} = C_x[i]$. If the swapping of a_e and C_x has an impact on the ith link set, the heuristic goes to either Step (4.5) or (4.6), depending on whether $C_x \in \underline{\underline{A}}$ or $C_x \in \underline{\underline{U}}$, which is checked in Step (4.4). In the case $C_x \in \underline{\underline{U}}$, the function addBit$(e, i)$ is called if $C_x[i] = 1$ and $a_{e,[i]} = 0$; otherwise, removeBit(i, e) is called if $C_x[i] = 0$ and $a_{e,[i]} = 1$. In the former case the ith bit is flipped from "0" to "1," hence a link is added to L_i; while in the latter, the ith bit is changed from "1" to "0," where a link is removed from L_i. If $C_x \in \underline{\underline{A}}$, let C_x be currently assigned to link f. The function add&removeBit(i, e, f) is called if $C_x[i] = 1$ and $a_{e,[i]} = 0$; otherwise, the function add&removeBit(i, f, e) is called (i.e., $C_x[i] = 0$ and $a_{e,[i]} = 1$).

Before addBit(i, e), removeBit(i, e), add&removeBit(i, e, f) are introduced, the attributes of network links should be defined first, which facilitate high computational efficiency in the cost reduction evaluation process for each link set.

The Attributes of Links

A link set may contain one or multiple isolated fragments, which are called the *components* of the link set. Each link of link set L_j could be attributed into one of the following four categories:

3.3 UFL for Multiple Failures

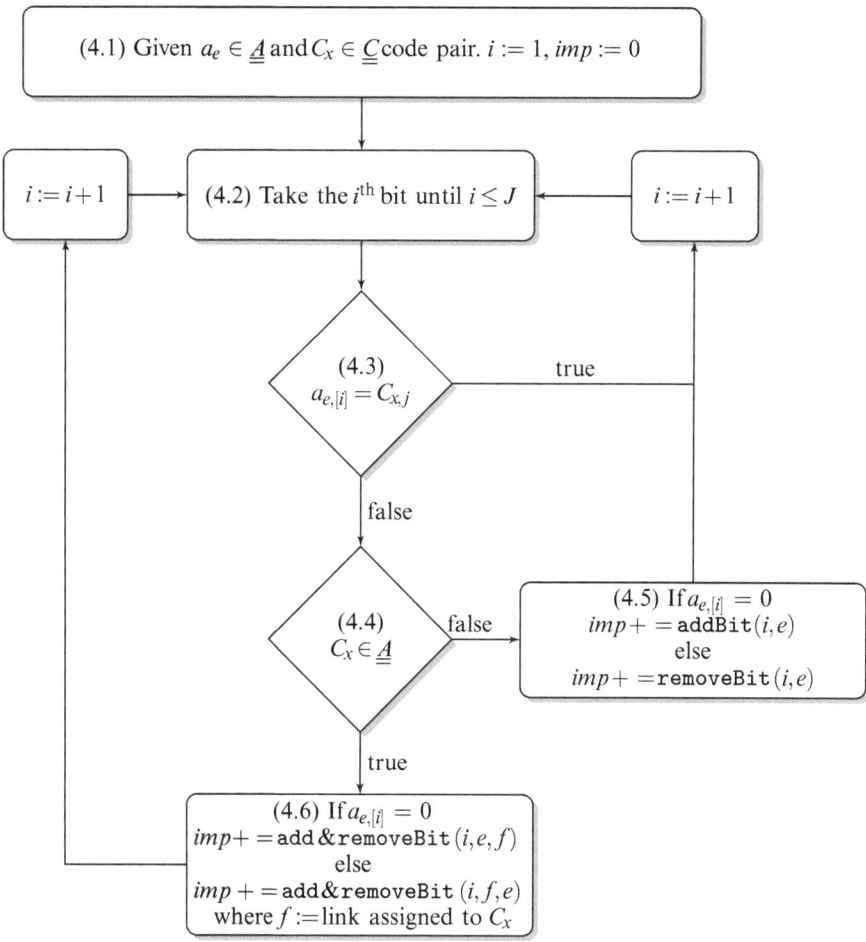

Fig. 3.22 Cost reduction evaluation for each code swapping

Isolated link   is a link not connected to any other link of the link set. Identifying these links is simple, since their both terminating nodes have degree 1.

Leaf link   is a link with exactly one of its terminating nodes of nodal degree 1, as shown in link (c, r) and (u, z), etc., in Fig. 3.23.

Bridge link   has both terminating nodes with a nodal degree larger than 1. Moreover, if the link is erased, then the component falls apart into two sub-components. To identify a bridge link, every 2-connected component must be identified first, which can be done in linear time [31]. For these links both terminating nodes of the link must belong to different 2-connected components. An example is (c, e) in Fig. 3.23.

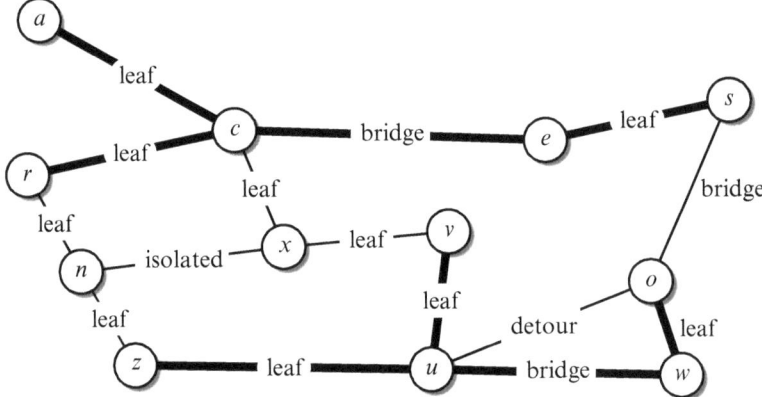

Fig. 3.23 An example on link attribute categorization for fast cost reduction evaluation on code swapping. The links of L_j are drawn with *thick lines*. The category of a link $e \in E \setminus L_j$, the *thin links*, is understood with respect to the graph given by *the thick links and e*

Table 3.6 Table of return values for function addBit(i, e)

Attribute of e for L_i	# of m-trails
Isolated	Increased by 1
Leaf	Unchanged
Bridge	Decreased by 1
Detour	Unchanged

Detour link is a part of a component and removal of it does not tear the component apart. For these links both terminating nodes of the link must belong to the same 2-connected component. An example is (o, u) in Fig. 3.23.

Next, the edges not part of the link set L_j are also assigned to one of these four categories. The category of a link $e \in E \setminus L_j$ is understood with respect to the graph given by $L_j \cup \{e\}$. For example, link (x, n) in Fig. 3.23 would be an isolated link of $L_j \cup \{(x, n)\}$.

addBit(i, e) returns the cost reduction in case $a_{e,[i]} = 0$ and $C_x[i] = 1$. In this case, because the ith bit of the two link codes is changed from "0" to "1," the cover length of the resultant m-trail solution will be increased by 1, while the number of m-trails could be increased or reduced or unchanged according to the attribute of link e with respect to the link set L_i. Table 3.6 summarizes the link attribute categorization.

removeBit(i, e) returns the cost reduction in case $a_{e,[i]} = 1$ and $C_x[i] = 0$. Because the ith bit is changed from "1" to "0," the cover length is decreased by 1, while the number of bm-trails should be updated according to the attribute of e with respect to L_i. This is summarized in Table 3.7.

add&removeBit(i, e, f) is for the cost reduction evaluation in the event that link e is added and another link f is removed from L_i. After the swapping the

3.3 UFL for Multiple Failures

Table 3.7 Table of return values for function `removeBit(i, e)`

Attribute of e for L_i	# of bm-trails
Isolated	Decreased by 1
Leaf	Unchanged
Bridge	Increased by 1
Detour	Unchanged

Table 3.8 Table of return values for function `add&removeBit(i, e, f)`

Add e Attrib. of L_i	Remove f Attrib. of L_i	# of bm-trails
Isolated	Isolated	**Unchanged**
	Leaf	**Increased by 1**
	Bridge	**Increased by 2**
	Detour	**Increased by 1**
Leaf	Isolated	**Decreased by 1** if e and f are not adjacent links, otherwise **unchanged**
	Leaf	**Either unchanged or increased by 1**, if e is connected to f. See link (r, n) and (r, c) on Fig. 3.23 as an example.
	Bridge	**Increased by 1**
	Detour	**Unchanged**
Bridge	Isolated	**Decreased by 2** if e and f are disjoint, otherwise **increased by 1**
	Leaf	Either **Decreased by 1**, or **unchanged** if e is adjacent to f. See link (o, w) and (o, s) on Fig. 3.23 as an example.
	Bridge	**Unchanged**
	Detour	**Decreased by 1**
Detour	Isolated	**Decreased by 1**
	Leaf	**Unchanged**
	Bridge	Either **Decreased by 1** or **unchanged** if e reconnects the detached subcomponents. See links (o, u) and (u, w) in Fig. 3.23 as an example. Lemma 3.6 provides a method to determine a bridge.
	Detour	**Unchanged**

cover length is unchanged while the number of m-trails changes according to the attributes of both links. This is provided in Table 3.8.

In summary, the proposed GCS swaps a code pair with the steepest cost reduction larger than a threshold γ based on the proposed link attribute categorization and table lookup process, which greedily approaches to better performance according to Eq. (3.1). With GCS, very high computational efficiency can be achieved thanks to the constant time complexity in evaluating each code pair, which will be detailed in the next subsection. The prototype of the proposed algorithm can be found in [35].

3.3.7.2 Computational Complexity Analysis

The cost reduction evaluation is performed in each code swapping, which dominates the computational complexity of the heuristic algorithm. Next we prove three claims needed for a lemma that describes the computational complexity of the cost reduction evaluation process for a single code swapping.

Lemma 3.6. *The complexity of* CostREval(i, j) *is* $O(1)$.

Proof. In the function CostREval(i, j) a lookup in one of the Table 3.6, 3.7 or 3.8 is performed depending on to the categories of the links corresponding to codes of i and j. We need to show that every table entity can be evaluated in constant time.

The execution of add&removeBit(i, e, f) with e and f as a detour and bridge link, respectively, is the only not obvious case. Here, for a link set L_j a bridge link $f \in L_j$ and for two nodes u, v, where $e = (u, v)$, we have to determine whether u and w are in the same component of $L_j \setminus \{e\}$. Such a function can be implemented by storing the *reach* and *leave* order of each node in the DFS traversal. It can be performed in the pre-calculation process beside the link attribute categorization in Step (2). As shown in Fig. 3.24 as an example, the reach order of the DFS is written on the top of nodes, while the leave order is below the nodes. Let I_R and I_L denote the largest reach and the smallest leave order indices of the bridge. Every node with a reach and leave index at least I_R and at most I_L belongs to one side of bridge. As exemplified in Fig. 3.24, we have $I_R = 2$ and $I_L = 8$, thus $\{a, r, c\}$ are in one sub-component.

Lemma 3.7. *The complexity of Step (2) is* $O(|E|d^2 \log |E|)$.

Proof. Clearly, a DFS function for detecting the components in each link set is in $O(|E|)$ time complexity. Since each bridge link connects to two 2-connected components, to identify a bridge link, we must identify the corresponding two 2-connected components first, which can be done in $O(|E|)$ time complexity [31], for $|E| \geq |V|$. Also, such a check needs to go through each bit position $1, \ldots, J$, which multiplies the complexity by $J = O(d^2 \cdot \log |E|)$ according to the CGT construction [19]. Thus, the lemma is proved.

Lemma 3.8. *The complexity of Step (3) is* $O(|E|^2 d^2 \log |E|)$.

Proof. For each code and link pair, the proposed method can evaluate the possible cost reduction in a constant time ($O(1)$) according to Lemma 3.6. Since the overall

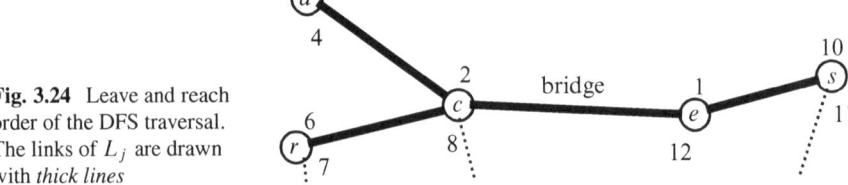

Fig. 3.24 Leave and reach order of the DFS traversal. The links of L_j are drawn with *thick lines*

time complexity is $O(|E| \cdot \#codes \cdot J)$, and since $\#codes = O(|E|)$, we have the worst case complexity of Step (3) as $O(|E|^2 d^2 \log |E|)$.

Lemma 3.9. *The computational complexity of a cost reduction evaluation process for a single swapping is $O(|E|^2 d^2 \log |E|)$.*

Proof. It is a direct consequence of the above three lemmas.

It is clear that the number of isolated components (or m-trails) in the initial RCA for each bit position cannot be more than $|V|/2$. This is because each isolated component consists of at least a single edge and two nodes, where $|V|$ is the number of nodes in the network. After each loop (defined in Steps (3), (4), (5), and (6) of Fig. 3.3), at least one m-trail is determined and erased from the link set; thus, the maximum number of code swapping should be upper bounded by $\frac{|V|}{2} \cdot J$, where J is the code length (in bits). Note that J is in the order of $O(d^2 \cdot \log |E|)$ according to the CGT construction [19]. By considering the complexity of each code swapping as $O(|E|^2 \log |E|)$, the overall worst case complexity in the proposed method is $O(|V||E|^2 \cdot d^2 \cdot \log^2 |E|)$. Compared with the scheme in [2, 3] with a complexity of $O(|E|^{2d} \cdot |V|^2)$, the proposed approach can achieve much better efficiency.

3.4 Performance Evaluation of UFL via a Central Controller

Extensive simulation on thousands of different random topologies was conducted to verify the proposed algorithms.

3.4.1 Performance Evaluation of RCA–RCS for Single-Link UFL

In this section we investigate the impacts of topology diversity on m-trail solutions for single-link UFL.

3.4.1.1 Topology Diversity on M-Trail Solutions

This section demonstrates the impact on m-trail solutions due to topology diversity. A huge number of experiments on thousands of randomly generated topologies were conducted.

Figure 3.25 shows the minimal number of m-trails versus network density on topologies with 50 nodes. The network connectivity was increased starting from a backbone ring with 50 nodes and 50 links to a fully meshed topology with 50 nodes and 1,225 links, where one or a few links were randomly added to the topology for each data set. To make it statistically meaningful, every data interval in Fig. 3.25 is 95 % confidence interval around the mean of 20 different topologies each obtained by randomly adding the same number of links to the backbone ring. We observed that the normalized length of the alarm code (i.e., from the length of alarm code we subtracted $\lceil \log_2 (|E| + 1) \rceil$) dramatically goes down when we add 50 to 100 links, or increase the nodal degrees from 2 to 3. See Fig. 3.25a, b, respectively. The length of the alarm code approaches the lower bound (i.e., $\lceil \log_2 (|E| + 1) \rceil$) when γ is large enough in Eq. (3.1). From Fig. 3.25a, b, we have also observed that when γ is small, the confidence interval for each data is larger than in the case of larger γ. This indicates the fact that both monitoring cost and bandwidth cost are more sensitive to different amounts of degree-2 nodes in the network, possibly due to the interplay between the two objectives in the cost function of Eq. (3.1).

Figure 3.25c shows the normalized cover ratio (i.e., the sum of cover length of all m-trails divided by $|E|$) versus average nodal degree. We have seen that the cover ratio slightly increases when the average nodal degree is increased from 2 to 9 for all the three γ values. Particularly, the cases with $\gamma = 5$ and 10 have better suppressed the increase of cover length ratio, which demonstrates the effectiveness in the tradeoffs between the length of the alarm code and bandwidth cost by manipulating the cost ratio γ.

Figure 3.26 shows the m-trail solutions with different numbers of nodes in the network topologies. Figure 3.26a, b shows the length of the alarm code versus the number of links. Clearly, when $\gamma = 1{,}000$, the length of the alarm code is only affected by the number of links when it is small, while quickly converges to $\lceil \log_2(|E| + 1) \rceil$ when the number of added links is increasing, regardless the number of nodes in the topology. Moreover, the length of alarm code is always $\lceil \log_2(|E| + 1) \rceil$ in case the network does not contain any node with a nodal degree 2 or smaller, which also verifies the observations. On the other hand, when γ is small, the convergence becomes slower, and the length of alarm code deviates more from $\lceil \log_2(|E| + 1) \rceil$ as γ is reduced.

Figure 3.26c shows the results of the experiment, on the relationship between normalized length of the alarm code and total number of degree-2 nodes in the topologies with 20, 30, 40, 50, and 60 nodes, altogether resulting 5,320 different random topologies. γ was set to 1,000. Each data point was obtained by averaging the results on 20 randomly generated topologies of the same total number of degree-2 nodes. We observed that all the topologies with different numbers of nodes require more m-trails for UFL as the number of degree-2 nodes grows, which meets our expectation. Interestingly, it is observed that all the topologies have to take similar normalized length of alarm code for UFL given the same number of degree-2 nodes. This indicates a serious impact on m-trail solutions due to the number of degree-2 nodes in a topology. Based on these experiments the following relationship can be taken as a rule of thumb for approximating the normalized length of the alarm code in a random network topology when γ is very large:

$$b = \text{\#m-trails} \approx \lceil \log_2(|E| + 1) \rceil + \frac{\text{\#degree-2 nodes}}{2} \quad (3.37)$$

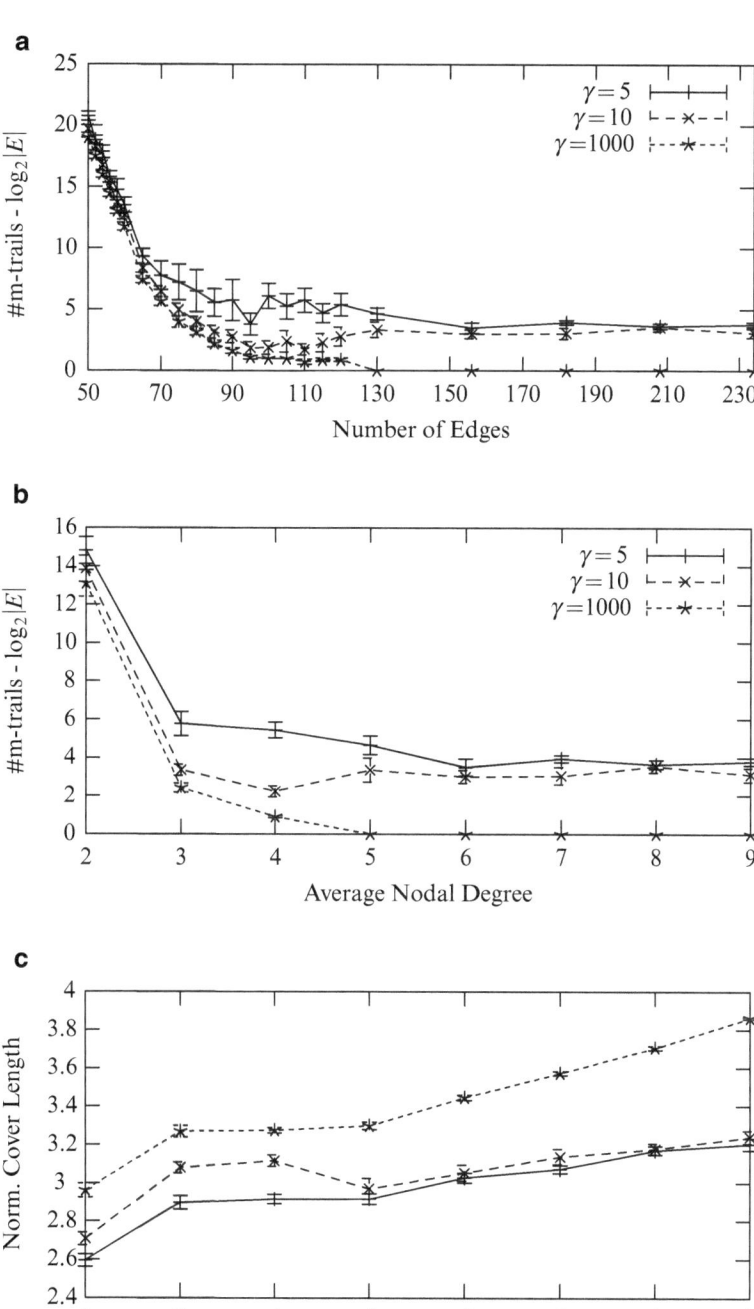

Fig. 3.25 Simulation on random topologies with 50 nodes. (**a**) Normalized length of alarm code versus number of links. (**b**) Normalized length of alarm code versus the average nodal degree. (**c**) Normalized cover ratio versus average nodal degree

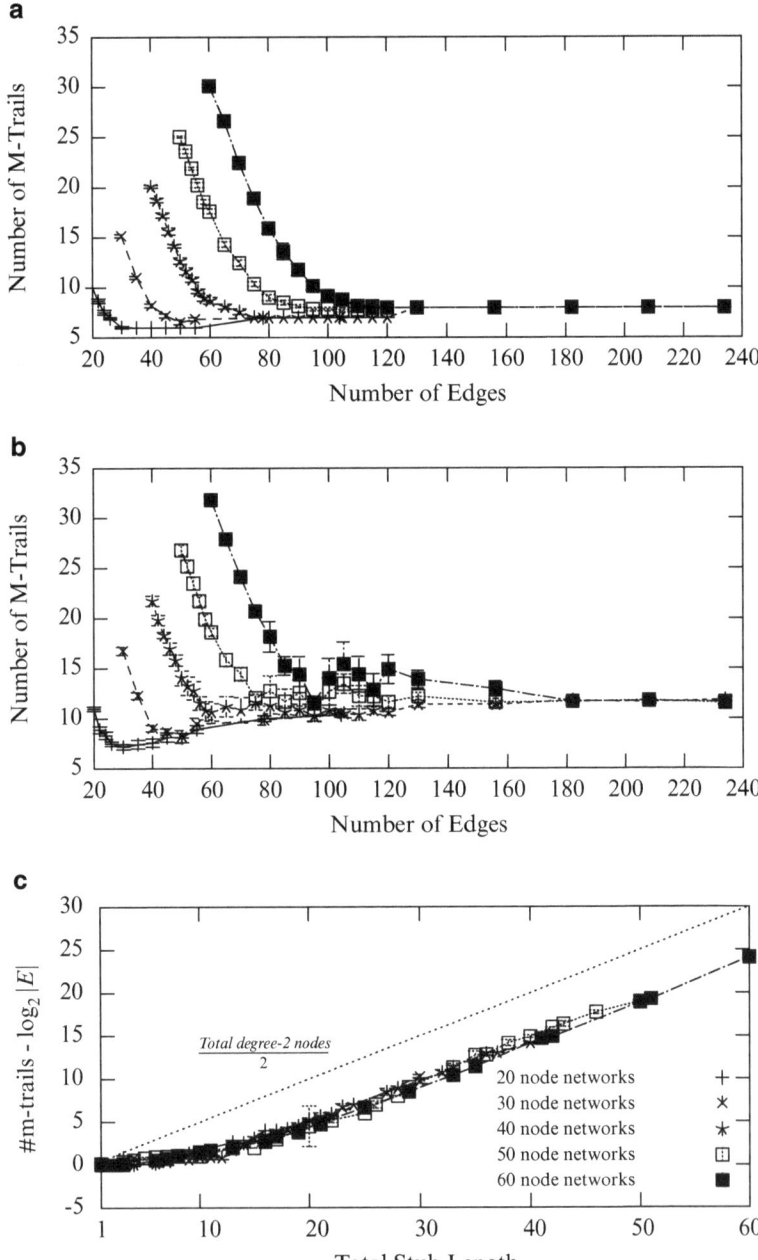

Fig. 3.26 The length of alarm code versus the number of added links for topologies of 20, 30, 40, 50, and 60 nodes. (**a**) $\gamma = 1000$, (**b**) $\gamma = 5$, (**c**) normalized length of alarm code versus total degree-2 nodes

Note that the above rule of thumb does not hold for every graph because the construction in Sect. 3.2.2.2 is a counterexample.

3.4.2 Performance Evaluation of AFL and LCC for Sparse-SRLG UFL

Here we compare the performance of different UFL approaches proposed for sparse-SRLG failures. The AFL corresponds to the scheme in Sect. 3.3.5; CA corresponds to the cycle accumulation method in [1], which allocates monitoring lightpaths in a shape of cycle one by one using Suurballe's algorithm to distinguish each SRLG pair; LCC corresponds to the method in Sect. 3.3.6 and CGT-GCS^3 is based on $\overline{3}$-separable constructions introduced in Sect. 3.3.7, which was originally designed for the dense-SRLG scenario.

3.4.2.1 Adjacent Dual-Link Failure Scenario

The motivation behind the adjacent dual-link failure scenario is that the geographically adjacent links incident to a common node are very likely put into a conduit for some distance and exposed to a common risk of being cut [9, 25].

The result on the minimum number of m-trails required in the scenario of SRLGs containing single failures and 10 % of adjacent dual-link failures is shown in Fig. 3.27. First, we find that the number of m-trails increases when the network size grows, which meets our expectation. In particular, the AFL can achieve superb scalability due to the following two aspects. First, compared with CA and CGT-GCS^3, AFL based on m-trails can explore the largest problem design space in terms of network topology diversity. This is attested in Fig. 3.27, where CA has achieved worse performance than AFL when network is dense (i.e., girth $g = 3$). We have also seen that when network is getting more sparsely connected (i.e., in the case of larger values of g), the performance advantage of the AFL algorithm grows because CGT-GCS^3 was not designed for sparse SRLGs. Note that with AFL, the number of m-trails can be well upper bounded by the number of links in the topology. Secondly, the AFL algorithm can take advantage of the precise SRLG information on link code dependency, which is nonetheless absent in the design of CA and CGT-GCS^3. Thus, AFL can achieve much better performance, especially when the SRLGs are sparse and heterogeneous in terms of the number of links contained in each SRLG.

It is observed in Fig. 3.28 that both AFL and LCC achieve much better performance than link based monitoring, and LCC slightly outperforms AFL as the permissive rule allows more general coding patterns than the strict rule (shown in Fig. 3.16), on which basis AFL was designed.

The scenarios of larger SRLG sets are simulated (50 % dual adjacent failures), and the results are given in Fig. 3.29. Firstly we find that AFL outperforms CA in

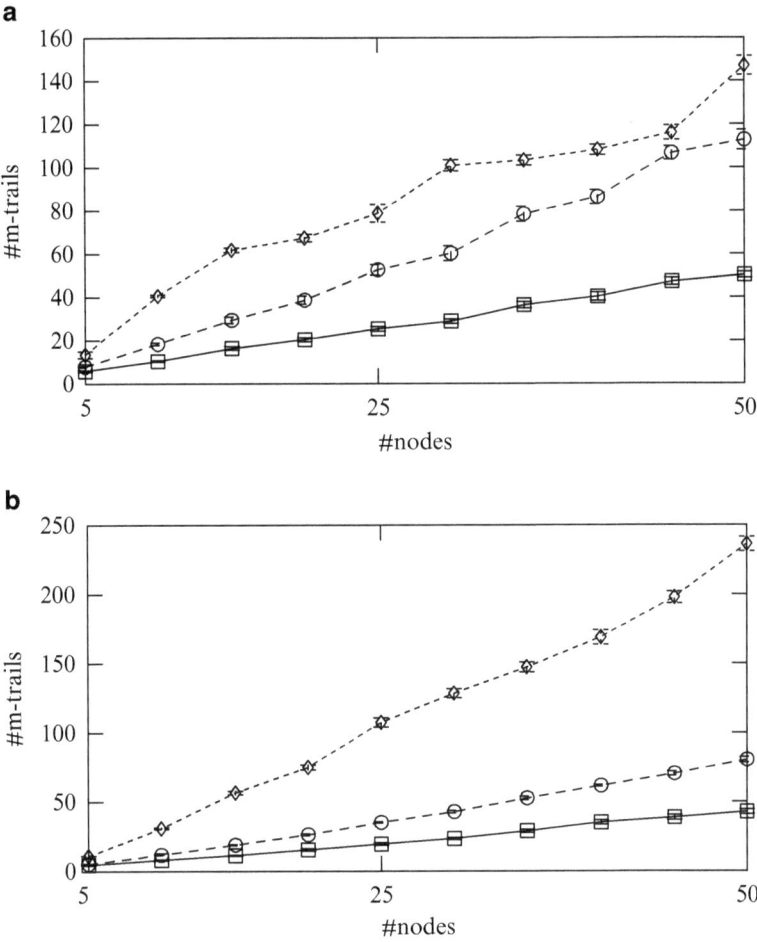

Fig. 3.27 The number of m-trails versus the number of nodes with 10 % of adjacent dual SRLGs and different girth parameters $g = 3$ and 7, where AFL, CA and CGT-GCS^3 is denoted by *open square*, *open circle*, and *open diamond*, respectively. (**a**) Dense topologies ($g = 3$), (**b**) sparse topologies ($g = 7$)

terms of the number of m-trails in both cases, and the advantage grows when more multi-link SRLGs exist in the network. This is due to a more flexible monitoring structure (which is a free-routed non-simple trail) employed in the proposed AFL. Note that with CA, the routing of the monitoring cycles relies on Suurballe's algorithm and least-cost in nature, which gives results far from the logarithmic bound on the number of monitoring cycles. Also from Fig. 3.27 we find that due to the consideration of sparse SRLGs, AFL and CA achieved much better performance than CGT-GSC^3.

3.4 Performance Evaluation of UFL via a Central Controller

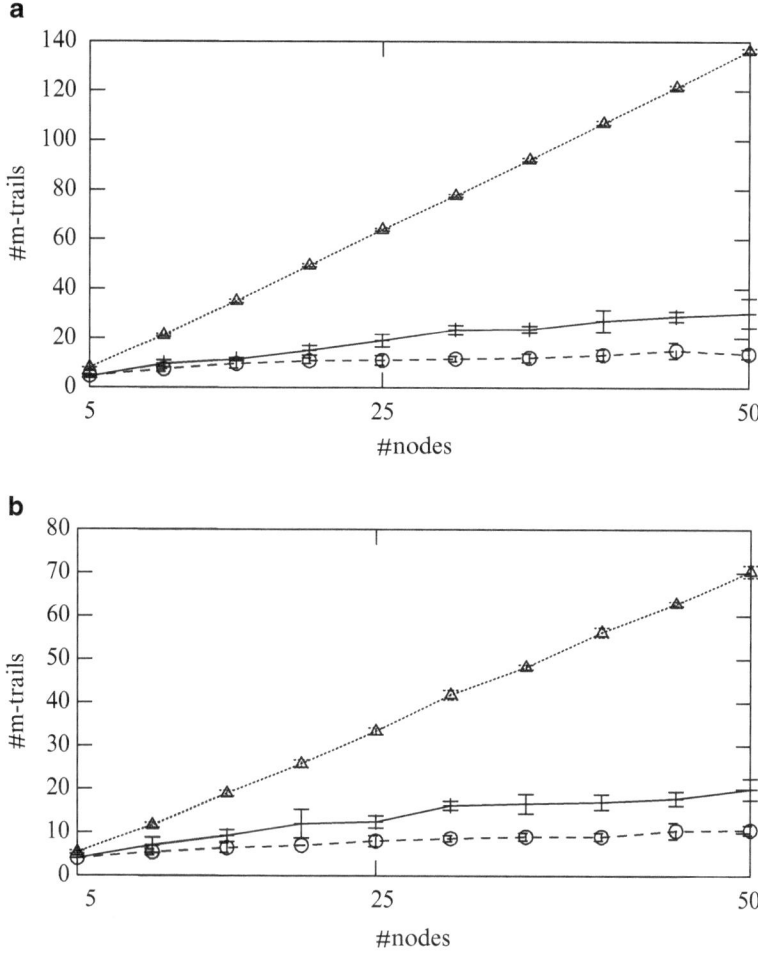

Fig. 3.28 The number of bm-trails versus the number of nodes with 10 % of adjacent dual SRLGs, where LCC, AFL, and link-based monitoring is denoted by *open circle*, *plus symbol*, and *open triangle*, respectively. (**a**) Dense topologies ($g = 3$), (**b**) sparse topologies ($g = 7$)

3.4.2.2 Single Node Failure

Because of the strict rule on the link codes shown in Fig. 3.16a, AFL approaches link-based monitoring in this scenario. However, LCC heuristic built on the permissive rule on the link codes shown in Fig. 3.16b perfectly fits into this scenario. The results are shown in Fig. 3.30, which clearly shows the superior performance of LCC in terms of the number of required bm-trails for node UFL.

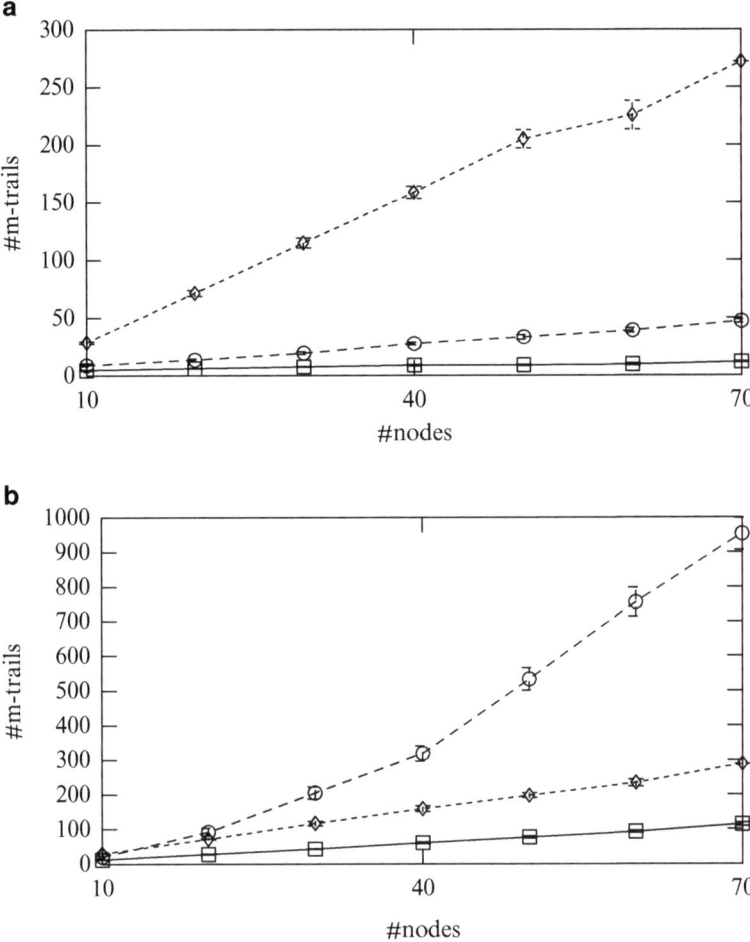

Fig. 3.29 The number of m-trails versus the number of nodes with different SRLG levels, with girth parameter $g = 5$, where AFL, CA and CGT-GCS^3 is denoted by *open square*, *open circle* and *open diamond*, respectively. (**a**) Single-link failures, (**b**) sparse SRLG failures

3.4.3 Performance Evaluation of CGT-GCS for Dense-SRLG UFL

In the simulation of multiple link failures (dense-SRLGs), the scheme introduced in Sect. 3.3.7 is denoted as $CGT\text{-}GCS^1$, $CGT\text{-}GCS^2$, and $CGT\text{-}GCS^3$ for failure localization of SRLGs with up to 1, 2, and 3 links, respectively, where CGT codes based on $\overline{d}$-separable constructions with $d = 1$, $d = 2$, and $d = 3$ are employed. CA^1 and CA^2 corresponds to the method that each bidirectional m-trail is allocated one after the other to distinguish each pair of SRLGs using any

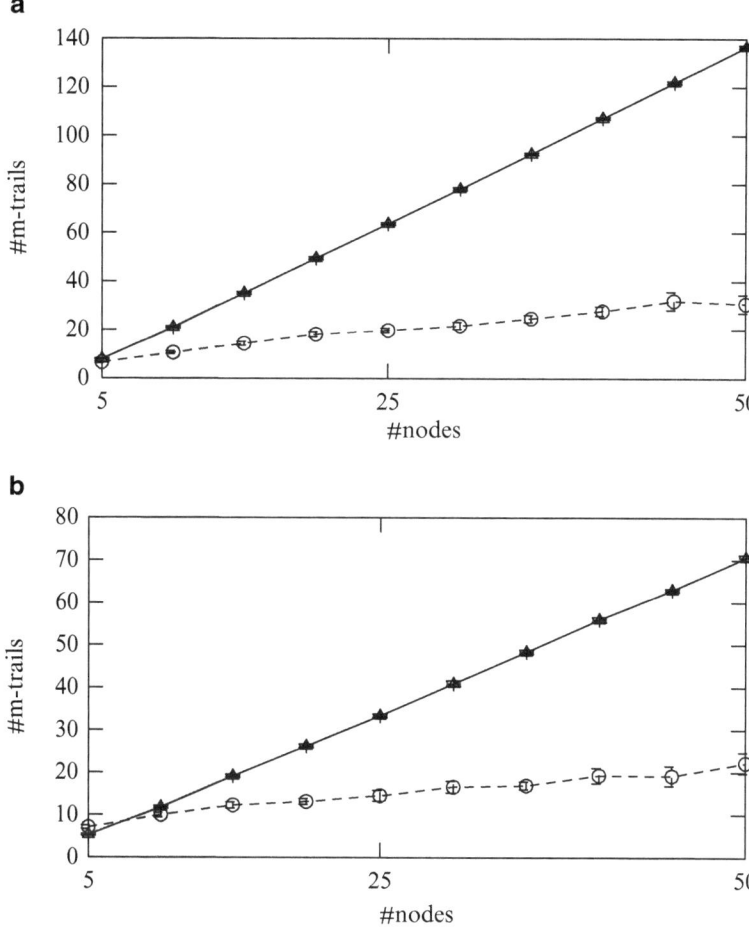

Fig. 3.30 The number of bm-trails versus the number of nodes with *all single-link and node failures*, where LCC, AFL, and link-based monitoring is denoted by *open circle, plus symbol,* and *open triangle*, respectively. (**a**) Dense topologies ($g = 3$), (**b**) sparse topologies ($g = 7$)

Dijkstra's algorithm based scheme, such that UFL for single-link SRLGs and for both single- and double-link SRLGs can be achieved, respectively. The method is generic and has been considered in a number of previously reported studies [2,5,52]. We have also implemented the construction in [18] that used disjoint spanning trees, denoted as $DSTC$. The construction provides an upper bound for $(d+1)$-connected topologies, but is invalid for topologies with any node of a smaller nodal degree than $(d + 1)$. The upper bound is given by $(d + 1) \cdot J$, where J is the length of the CGT codes employed.

Figure 3.31b shows the lengths of CGT codes (i.e., J) versus the number of links $|E|$ of the corresponding topology by the CGT code generator GEN_CGT in [14],

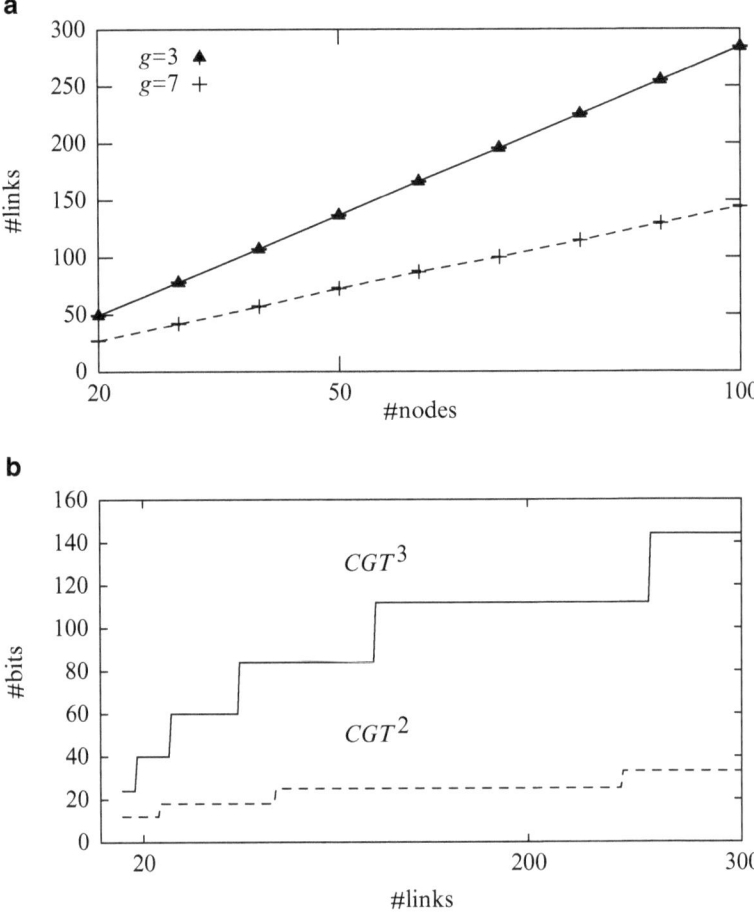

Fig. 3.31 Statistics on the input data. (**a**) Statistics of the random topologies. The dense networks have girth $g = 3$, while for the sparse networks $g = 7$. (**b**) The length of CGT code (in bits) versus the number of links as an input to the CGT code generator GEN_CGT in [14]

where the scenarios with $d = 2$ are presented. It is intuitive that when SRLGs with more links are considered, longer CGT codes are required for each link.

The performance metrics employed in the comparison of the six schemes are the minimum number of m-trails required for achieve UFL. Both metrics are examined with respect to different network sizes (i.e., the number of nodes) and topology densities (i.e., g values), which will be presented in the following two subsections. The simulation has been done on over 800 randomly generated topologies, and each data point was obtained by averaging the results from 10 different topologies with a specific g value and number of nodes. A bar for each data in the charts shows the 95 % confidence interval.

3.4 Performance Evaluation of UFL via a Central Controller 111

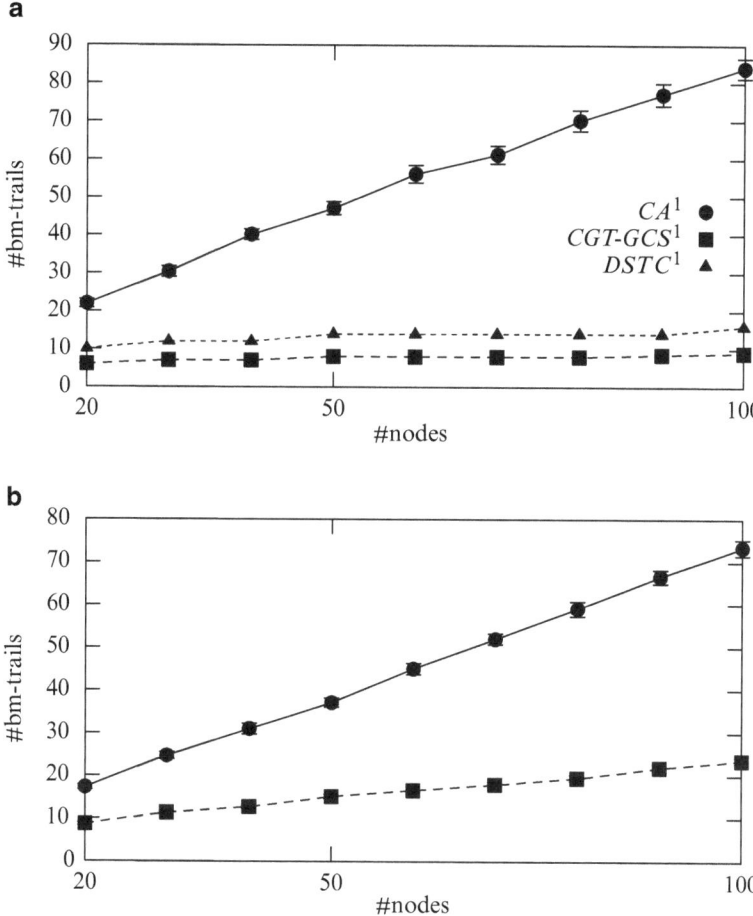

Fig. 3.32 The number of bm-trails versus the number of nodes for *single* failures. (**a**) $g = 3$ single failure, (**b**) $g = 7$ single failure

3.4.3.1 Number of M-Trails Versus Network Size

The performance in terms of the minimum number of m-trails is first investigated, and the results are shown in Figs. 3.32, 3.33, and 3.34. First, we find that the number of m-trails increases when the network size grows, which is observed in all the cases. It clearly shows that the proposed approach achieves much better scalability, where $CGT\text{-}GCS^1$, CA^1, and CA^2 have achieved far worse performance than $CGT\text{-}GCS^1$ and $CGT\text{-}GCS^2$, respectively, in both types of network topologies.

The superior performance of the proposed approach in minimizing the number of m-trails can be explained as follows. First, the m-trails have the most flexible routing structure that can fully explore the solution space. This serves as a critical

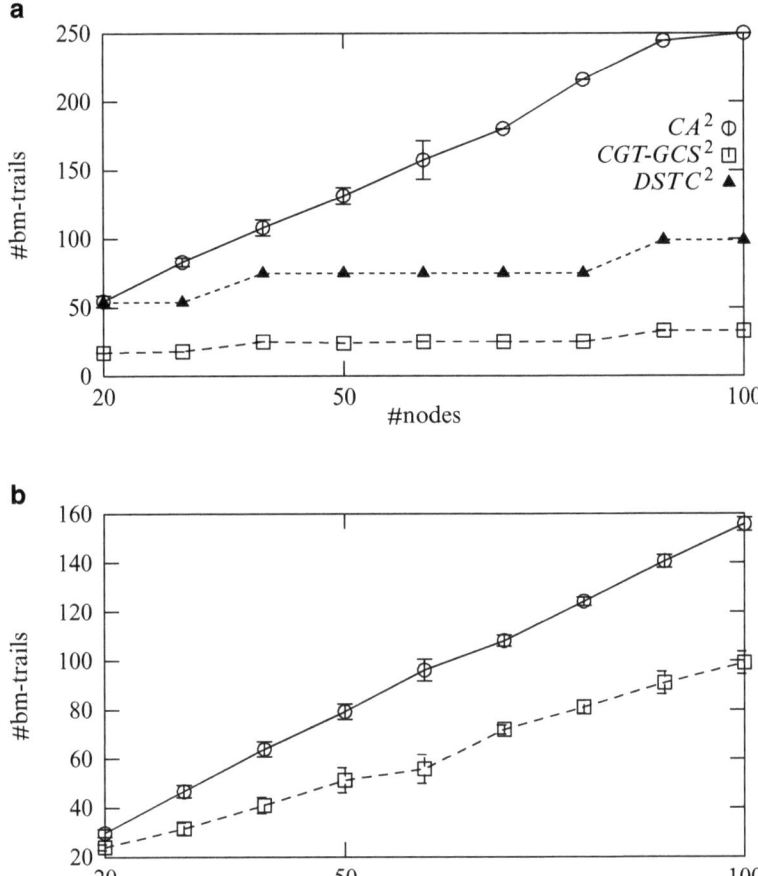

Fig. 3.33 The number of bm-trails versus the number of nodes for *double* failures. (**a**) $g = 3$ double failure, (**b**) $g = 7$ double failure

factor in overcoming the vicious effect of topology diversity. It can be attested that our scheme has less advantage against the other two counterparts when the network is sparsely connected (i.e., $g = 7$), because there are less alternatives in allocating the m-trails. Second, because CA^1 and CA^2 have each monitoring lightpath sequentially allocated into the network using a shortest path routing algorithm, they lack intelligence in exploring the design space and network topology diversity. Third, $DSTC$ [18] tries to ensure the code uniqueness of each SRLG by using $d + 1$ edge-disjoint spanning trees, which is strongly limited by the topology connectivity. It is clearly shown that the construction can only yield valid solution in very densely meshed topologies, while fails in most of the sparse topologies considered in the simulation.

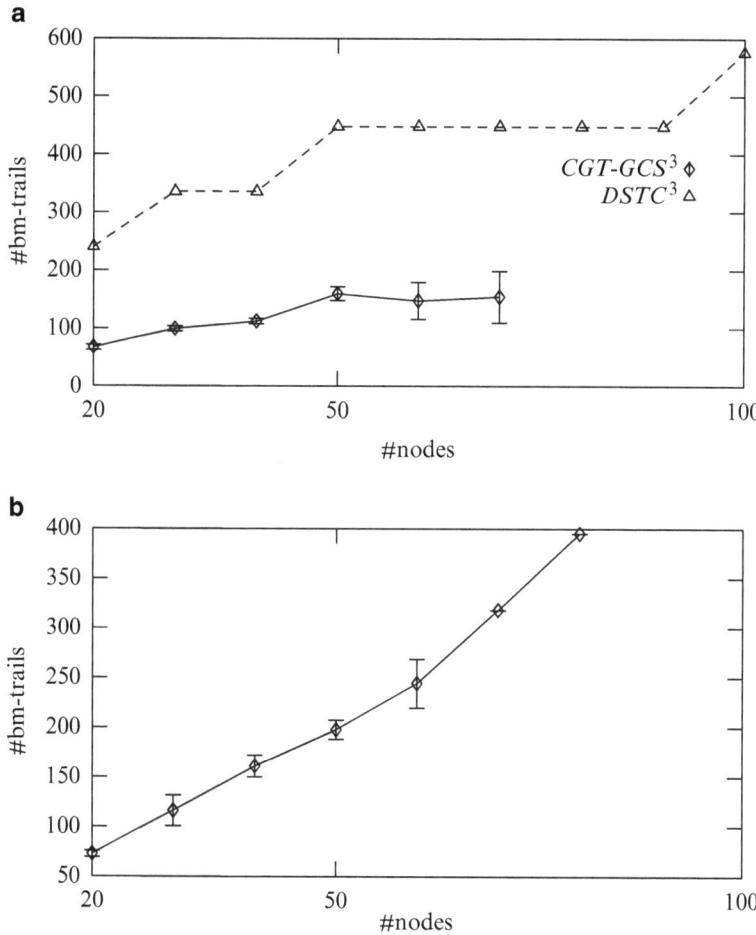

Fig. 3.34 The number of bm-trails versus the number of nodes for *triple* failures. (**a**) $g = 3$ triple failure, (**b**) $g = 7$ triple failure

3.5 Summary

In this chapter, we explored the theoretical background of UFL via a central controller. Lower bounds on the number of (b)m-trails were presented for several graph topologies. Besides single-link failures the problem of multiple link failures was discussed, and algorithms for the sparse- and dense-SRLG scenario were introduced. The design principles and the characteristics of these approaches were discussed. In order to demonstrate the efficiency of these methods simulation results were provided. Although theoretically sound and failure localization via a central controller gives us a solid background, it still cannot eliminate signaling from the failure localization and failure notification processes. Thus, further elaborations required to reach all-optical signaling-free restoration. A step towards this goal is made in Chap. 4, where distributed failure localization is discussed, which manages to get rid of any control plane signaling in the failure localization phase.

References

1. Ahuja S, Ramasubramanian S, Krunz M (2008) SRLG failure localization in all-optical networks using monitoring cycles and paths. In: Proceedings of the IEEE INFOCOM, pp 181–185
2. Ahuja S, Ramasubramanian S, Krunz M (2009) Single link failure detection in all-optical networks using monitoring cycles and paths. IEEE/ACM Trans Netw 17(4):1080–1093
3. Ahuja S, Ramasubramanian S, Krunz M (2011) SRLG failure localization in optical networks. IEEE/ACM Trans Netw 19(4):989–999
4. Alspach B (2008) The wonderful Walecki construction. Bull Inst Combin Appl 52:7–20
5. Assi C, Ye Y, Shami A, Dixit S, Ali M (2002) A hybrid distributed fault-management protocol for combating single-fiber failures in mesh based DWDM optical networks. In: Proceedings of the IEEE GLOBECOM, pp 2676–2680
6. Babarczi P (2012) Survivable optical network design with unambiguous shared risk link group failure localization. Ph.D. Dissertation, Budapest University of Technology and Economics. http://lendulet.tmit.bme.hu/~babarczi/dissertation/Babarczi_PhD_Dissertation.pdf
7. Babarczi P, Tapolcai J, Ho PH (2011) Adjacent link failure localization with monitoring trails in all-optical mesh networks. IEEE/ACM Trans Netw 19(3):907–920
8. Babarczi P, Tapolcai J, Ho PH (2011) SRLG failure localization with monitoring trails in all-optical mesh networks. In: Proceedings of the international workshop on design of reliable communication networks (DRCN), Krakow, pp 188–195
9. Choi H, Subramaniam S, Choi H (2003) Loopback recovery from neighboring double-link failures in WDM mesh networks. Inf Sci 149(1–3):197–209
10. Demeester P, Gryseels M, Autenrieth A, Brianza C, Castagna L, Signorelli G et al. (1999) Resilience in multilayer networks. IEEE Commun Mag 37(8):70–76
11. Diestel R (2000) Graph theory. Springer, New York
12. Doumith EA, Zahr SA, Gagnaire M (2010) Monitoring-tree: an innovative technique for failure localization in WDM translucent networks. In: Proceedings of the IEEE GLOBECOM, pp 1–6
13. Du D, Hwang FK (2000) Combinatorial group testing and its applications. World Scientific, Singapore
14. Eppstein D, Goodrich M, Hirschberg D (2005) Improved combinatorial group testing for real-world problem sizes. In: Proceedings of the workshop on algorithms and data structures (WADS). Springer, Waterloo, pp 86–98
15. Gabow HN, Westermann HH (1992) Forests, frames, and games: algorithms for matroid sums and applications. Algorithmica 7(1):465–497
16. Garey M, Johnson D (1979) Computers and intractability. A guide to the theory of NP-completeness. A series of books in the mathematical sciences. WH Freeman, San Francisco
17. Haddad A, Doumith E, Gagnaire M (2013) A fast and accurate meta-heuristic for failure localization based on the monitoring trail concept. Telecommun Syst 52(2):813–824
18. Harvey N, Patrascu M, Wen Y, Yekhanin S, Chan V (2007) Non-adaptive fault diagnosis for all-optical networks via combinatorial group testing on graphs. In: Proceedings of the IEEE INFOCOM, pp 697–705
19. Hwang FK, Sós VT (1987) Non-adaptive hypergeometric group testing. Studia Sci Math Hungar 22:257–263

20. Li C, Ramaswami R, Center I, Heights Y (1997) Automatic fault detection, isolation, and recovery in transparentall-optical networks. IEEE/OSA J Lightwave Technol 15(10):1784–1793
21. Lidl R, Niederreiter H (1994) Introduction to finite fields and their applications. Cambridge University Press, Cambridge
22. Lucas E (1893) Recreations Mathematiques. Gauthier-Villars, Paris
23. Machuca C, Kiese M (2009) Optimal placement of monitoring equipment in transparent optical networks. In: Proceedings of the IEEE DRCN, pp 1–6
24. Maeda M (1998) Management and control of transparent optical networks. IEEE J Sel Areas Commun 16(7):1008–1023
25. Markopoulou A, Iannaccone G, Bhattacharyya S, Chuah C, Diot C (2004) Characterization of failures in an IP backbone. In: Proceedings of the IEEE INFOCOM, vol 4. Citeseer, pp 2307–2317
26. Mas C, Thiran P (2002) An efficient algorithm for locating soft and hard failures in WDM networks. IEEE J Sel Areas Commun 18(10):1900–1911
27. Mas C, Tomkos I, Tonguz O (2005) Failure location algorithm for transparent optical networks. IEEE J Sel Areas Commun 23(8):1508–1519
28. Moghaddam E, Tapolcai J, Mazroa D, Hosszu É (2011) Physical impairment of monitoring trails in all optical transparent networks. In: Proceedings of the international congress on ultra modern telecommunications and control systems and workshops (ICUMT). IEEE, Budapest, Hungary, pp 1–7
29. Nash-Williams C (1961) Edge-disjoint spanning trees of finite graphs. J Lond Math Soc 1(1):445–450
30. Rao N (1993) Computational complexity issues in operative diagnosis of graph-based systems. IEEE Trans Comput 42(4):447–457
31. Schrijver A (2003) Combinatorial optimization: polyhedra and efficiency. Springer, Berlin
32. Stanic S, Subramaniam S, Choi H, Sahin G, Choi H (2002) On monitoring transparent optical networks. In: Proceedings of the international conference on parallel processing workshops (ICPPW), pp 217–223
33. Stanic S, Subramaniam S, Sahin G, Choi H, Choi HA (2010) Active monitoring and alarm management for fault localization in transparent all-optical networks. IEEE Trans Netw Serv Manag 7(2):118–131
34. Suurballe JW (1974) Disjoint paths in a network. Networks 4:125–145
35. Tapolcai J, Babarczi P (2014) Demo web page on (b)m-trail design, http://lendulet.tmit.bme.hu/demo/mtrail
36. Tapolcai J, Ho PH, Rónyai L, Babarczi P, Wu B (2011) Failure localization for shared risk link groups in all-optical mesh networks using monitoring trails. IEEE/OSA J Lightwave Technol 29(10):1597–1606
37. Tapolcai J, Wu B, Ho PH, Rónyai L (2011) A novel approach for failure localization in all-optical mesh networks. IEEE/ACM Trans Netw 19(1):275–285
38. Tapolcai J, Ho PH, Rónyai L, Wu B (2012) Network-wide local unambiguous failure localization (NWL-UFL) via monitoring trails. IEEE/ACM Trans Netw 20(6):1762–1773
39. Tapolcai J, Ho PH, Babarczi P, Rónyai L (2013) On achieving all-optical failure restoration via monitoring trails. In: Proceedings of the IEEE INFOCOM, pp 380–384
40. Tapolcai J, Wu B, Ho PH (2009) On monitoring and failure localization in mesh all-optical networks. In: Proceedings of the IEEE INFOCOM, Rio de Janero, pp 1008–1016
41. Tutte W (1961) On the problem of decomposing a graph into n connected factors. J Lond Math Soc 1(1):221–230
42. Wen Y, Chan V, Zheng L (2005) Efficient fault-diagnosis algorithms for all-optical WDM networks with probabilistic link failures. IEEE/OSA J Lightwave Technol 23:3358–3371
43. Wu B, Ho PH, Yeung K (2008) Monitoring trail: a new paradigm for fast link failure localization in WDM mesh networks. In: Proceedings of the IEEE GLOBECOM
44. Wu B, Ho PH, Yeung K (2009) Monitoring trail: on fast link failure localization in all-optical WDM mesh networks. IEEE/OSA J Lightwave Technol 27(18):4175–4185

45. Wu B, Yeung K, Ho PH (2009) Monitoring cycle design for fast link failure localization in all-optical networks. IEEE/OSA J Lightwave Technol 27(10):1392–1401
46. Wu B, Ho PH, Tapolcai J, Babarczi P (2010) Optimal allocation of monitoring trails for fast SRLG failure localization in all-optical networks. In: Proceedings of the IEEE GLOBECOM
47. Wu B, Ho PH, Tapolcai J, Jiang X (2010) A novel framework of fast and unambiguous link failure localization via monitoring trails. In: Proceedings of the IEEE INFOCOM WIP, San Diego
48. Wu B, Ho PH, Yeung K, Tapolcai J, Mouftah H (2011) Optical layer monitoring schemes for fast link failure localization in all-optical networks. IEEE Commun Surv Tutor 13(1):114–125
49. Wu B, Yeung K, Hu B, Ho PH (2011) M^2-CYCLE: an optical layer algorithm for fast link failure detection in all-optical mesh networks. Elsevier Comput Netw 55(3):748–758
50. Zeng H, Huang C (2004) Fault detection and path performance monitoring in meshed all-optical networks. In: Proceedings of the IEEE GLOBECOM, vol 3, pp 2014–2018
51. Zeng H, Vukovic A (2007) The variant cycle-cover problem in fault detection and localization for mesh all-optical networks. Photonic Netw Commun 14(2):111–122
52. Zeng H, Huang C, Vukovic A (2006) A novel fault detection and localization scheme for mesh all-optical networks based on monitoring-cycles. Photonic Netw Commun 11(3):277–286
53. Zhao Y, Xu S, Wang X, Wang S (2010) A new heuristic for monitoring trail allocation in all-optical WDM networks. In: Proceedings of the IEEE GLOBECOM, pp 1–5
54. Zhao Y, Xu S, Wu B, Wang X, Wang S (2012) Monitoring trail allocation in all-optical networks with the random next hop policy. In: Proceedings of the IEEE high performance switching and routing (HPSR), pp 192–197

Chapter 4
Distributed Failure Localization

Abstract The chapter continues the topic of bm-trail allocation as in Chap. 3, by assuming a distributed control environment where a remote network controller for collecting the alarms is absent. Instead, the scenario that a node can individually perform UFL without relying on any failure notification mechanism is targeted. Accordingly, a constraint is imposed on the previously formulated bm-trail allocation problem where the alarms locally available to a node should form a complete alarm code table (ACT) for making the failure localization decision. This is also referred to as local unambiguous failure localization (L-UFL) at the node. A step further to L-UFL is that all the nodes are required to be L-UFL capable, which leads to the scenario referred to as Network wide L-UFL (NL-UFL). The chapter presents solutions to the bm-trail allocation problems for L-UFL and NL-UFL, respectively, via both bound analysis and heuristics under various network failure scenarios.

4.1 Introduction

Unambiguous Failure Localization (UFL) is defined in all-optical mesh WDM networks where any link failure can be precisely and instantly identified via monitoring a set of supervisory lightpaths. Such capability in the network optical layer is highly desired in order to meet the stringent requirement on service continuity and to support various failure-dependent restoration mechanisms [7, 10, 17].

Several UFL schemes have been proposed in Chap. 3. Using m-trails with bidirectional lightpaths (bm-trails) in all-optical WDM networks has been proposed [6, 9, 11, 14]; and all these studies set the destination node of each bm-trail as the only node that can detect the status of the bm-trail. As a consequence, alarm dissemination is needed upon a failure event mostly via flooding and dedicated failure notification based on network layer signaling, such that the alarm code can be formed at a remote site or any network node in order to localize the failure. Obviously, using electronic signaling in alarm bit collection incurs additional control complexity, operational overhead, and lower robustness of the system. The dependency on the upper layer signaling mechanism implies that these approaches leave the optical domain, which may cause problems. Such an issue was considered in [1], which aimed to minimize the number of monitoring locations in order to reduce the alarm dissemination. Nonetheless, the research in [1] can hardly be applicable to the scenario where all the nodes are required to perform UFL in the optical domain, and may yield solutions with multiple MLs in sparse networks, and in this case the MLs have to exchange their alarm bits via control plane signaling. Most importantly, the approach in [1] pays no attention to the number of required supervisory lightpaths which affects seriously the number of transmitters and the total cover length.

Motivated by the fact that UFL should be carried out completely in the optical domain and be free from any control plane signaling effort, this section investigates an advantageous scenario of all-optical failure localization using bidirectional m-trails. Here we define *Local UFL* (L-UFL) at a node if the node can individually perform UFL based on locally available on–off status of the traversing bm-trails; and *Network-Wide Local UFL* (NL-UFL) in the network if every node is L-UFL capable. By assuming that any node along a bm-trail can obtain the on–off status of the bm-trail via optical signal tapping, all the nodes traversed by the bm-trail can share the on–off status of the bm-trail. The proposed NL-UFL bm-trail allocation problem is to find a set of bm-trails such that every node can perform L-UFL based on the traversing bm-trails.

The section first introduces the NL-UFL framework along with its possible application scenarios, and defines the NL-UFL bm-trail allocation problem [13]. To gain an understanding of the problem, we conduct bound analysis by proving optimal solutions in terms of minimal cover length in a number of special graphs, such as lines, stars, and complete graphs. We also derive lower bounds for general graphs. Inspired by the optimal solution for complete graphs, we develop a novel heuristic algorithm for solving the NWL-UFL bm-trail allocation problem based on random spanning tree assignment (RSTA) and greedy link swapping (GLS). Extensive simulation on thousands of randomly generated topologies is conducted to verify the proposed heuristic approach and examine the derived lower bounds. The impact of network diversity is also investigated.

4.2 Problem Definition

4.2.1 Local Unambiguous Failure Localization

L-UFL is an advanced application of (b)m-trail deployment, where the UFL constraint is stricter while the shape constraint is the same as for UFL described in Sect. 3.1.2.1. A node is said to be *L-UFL capable* if and only if the node can perform UFL by inspecting the on–off status of the m-trails locally available. An m-trail is said to be *locally available* at node n if it terminates in n. In the L-UFL problem the goal is to find a node with L-UFL. This node is called *monitoring location (ML)*.

In many cases none of the nodes can serve as an ML, and thus the goal is to find a smallest set of nodes that altogether have sufficient information for UFL. As a consequence of Theorem 4.1, any node can be the single ML in a 2-connected

4.2 Problem Definition

network to achieve L-UFL with bm-trails for single-link failures. The bandwidth cost is also considered in the L-UFL problem, with significantly smaller weight compared to the number of MLs.

4.2.2 An L-UFL Example

Figure 4.1 shows an example of L-UFL m-trail solution on the SmallNet network for localizing any single-link failure. Node v_9 is chosen for monitoring location, and the local alarm code table (ACT) at node v_9 is also shown. Here, every m-cycle should terminate in node v_9.

4.2.3 State of the Art on L-UFL

The study in [1] set its goal to minimize the number of MLs, which is determined by analyzing the connectivity among the 2- and 3-edge-connected components in the topology. With each ML determined, a graph transformation is performed such that the MLs are merged into a supernode (denoted by m), and cycles are cumulatively added into the transformed graph one by one via Suurballe's algorithm. See also the example solution in Fig. 4.1 which is generated with this method. The method was extended for sparse SRLGs in [2].

4.2.4 Network-Wide L-UFL

In Network-wide L-UFL (NL-UFL) every node has to be L-UFL capable. To achieve this, the L-UFL condition is slightly relaxed by assuming that any node along a monitoring lightpath can obtain the on–off status of the lightpath via optical signal tapping. Thus, all the nodes traversed by the lightpath can share the on–off status of the bm-trail. It means a node can locally inspect the on–off status of every traversing monitoring lightpath.

The task is to allocate monitoring lightpaths with minimum total cover length, such that the shape constraint is fulfilled (see also Sect. 3.1.2.1), and the NL-UFL constraint requires unique alarm code for each SRLG at each node. The alarm code at node $v \in V$ for SRLG $z \in \mathscr{Z}$ is denoted by a_z^v, where $a_z^v = [a_{z,[1]}^v, a_{z,[2]}^v, \ldots, a_{z,[b^v]}^v]$ is a binary vector, b^v is the total number of bm-trails traversing node v, and the jth bit of a_z^v, denoted by $a_{z,[j]}^v$, is 0 if the j-th bm-trail traversing node v is operating after failure of z, and 1 otherwise.

Thus for each node v we define an ACT, denoted as $\underline{\underline{A}}^v$, which is a $|\mathscr{Z}| \times b^v$ size matrix with each row as a_z^v, $\forall z \in \mathscr{Z}$. Obviously $b \geq b^v$ for each $v \in v$, as

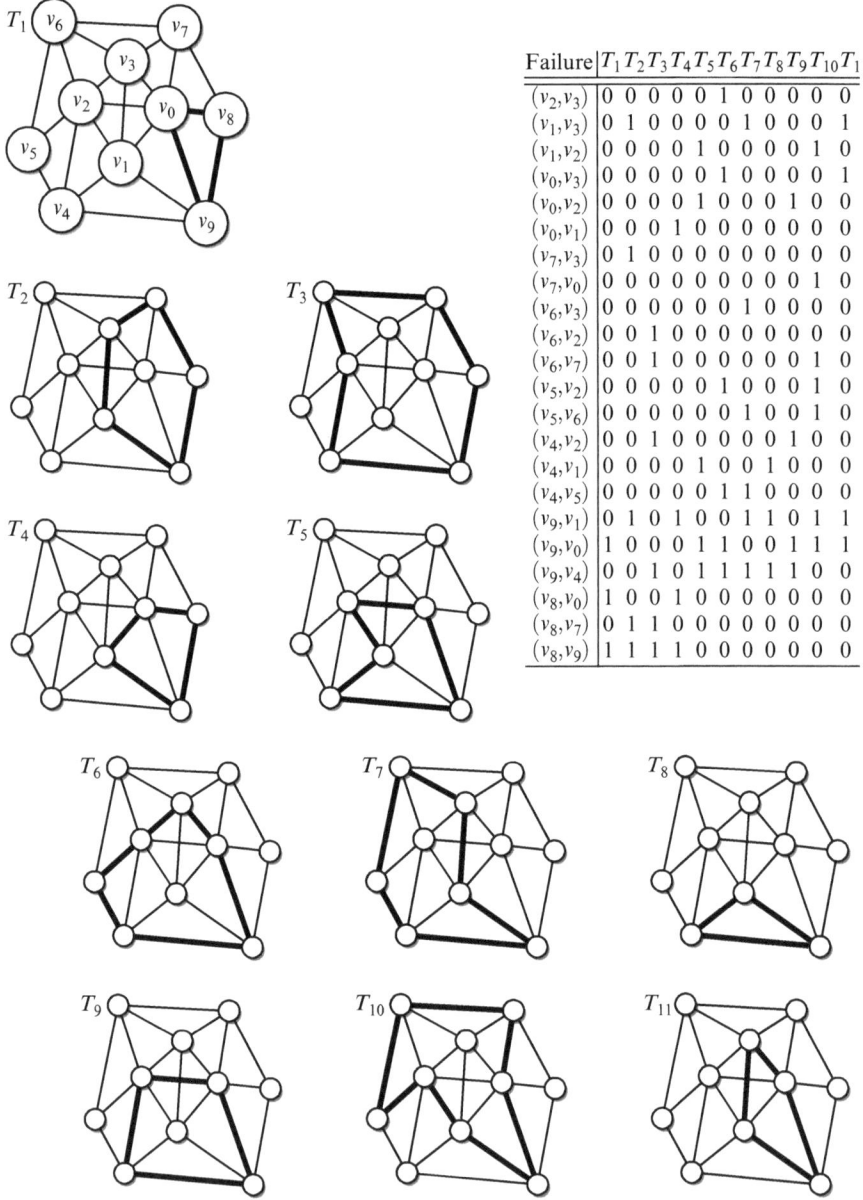

Fig. 4.1 Local unambiguous failure localization (L-UFL) based on m-cycles. Node 9 is the monitoring location. The cover length is $\|\mathcal{T}\| = 53$

ACTvv is a submatrix of the global ACT. When any failure occurs and interrupts one or a number of m-trails, node v will obtain a nonzero alarm code which uniquely identifies the failed SRLG.

The target function of the NL-UFL problem is the total cover length, denoted by $\|\mathcal{T}\|$. It is because NL-UFL does not rely on any upper layer signaling effort for alarm code dissemination and the length of alarm code is not as critical as that in UFL framework.

The following theorem indicates the feasibility of the approach in any 2-connected graph. The theorem demonstrates that an m-trail solution for single-link NL-UFL can always be obtained in a connected graph.

Theorem 4.1. *Given a connected graph, an m-trail solution for NL-UFL can always be found.*

Proof. The necessity of the connectivity of the graph is trivial. For the sufficiency part, we need to show that every node v_s can achieve L-UFL. Formally, we have to prove that the failure of an arbitrary link (v_i, v_j) can be unambiguously detected at v_s using a suitable collection of m-trails.

Indeed, after possibly renumbering the vertices we may assume that v_i is not farther from v_s than v_j. Then let T_1 be a (possibly empty) shortest path from v_s to v_i. Note that T_1 does not contain the link (v_i, v_j). Also, let $T_2 = T_1 \cup \{(v_i, v_j)\}$. Then we have that (v_i, v_j) is faulty iff T_2 gives an alarm signal and T_1 does not. These events are obviously visible at v_s and can be repeated for every link (v_i, v_j).

4.2.5 An NL-UFL Example

The L-UFL solution of Fig. 4.1 is not network wide as v_6 is traversed only by T_3, T_7, and T_{10}, and thus the ACT of node v_6 has only the corresponding three columns. This is not a valid ACT because (v_2, v_3) and (v_0, v_3) is not passed by any of these trails. Figure 4.2 shows an example of NL-UFL for any single-link SRLG using six bm-trails for the SmallNet topology. It shows the routes of the six bm-trails $T_1, \ldots, T_6$. In our case, each of the ten nodes can achieve single-link L-UFL by inspecting the locally available on–off status of the traversing bm-trails. Every bm-trail is a spanning tree, thus every node can inspect every bm-trail.

4.3 Bounds on Bandwidth Cost

This section presents our bound analysis for cover length in the proposed NL-UFL bm-trail problem. We will first look into a lower bound on general connected graphs, followed by the optimal solutions for a number of special graph topologies: line, star, and complete graphs. The results are based on [15].

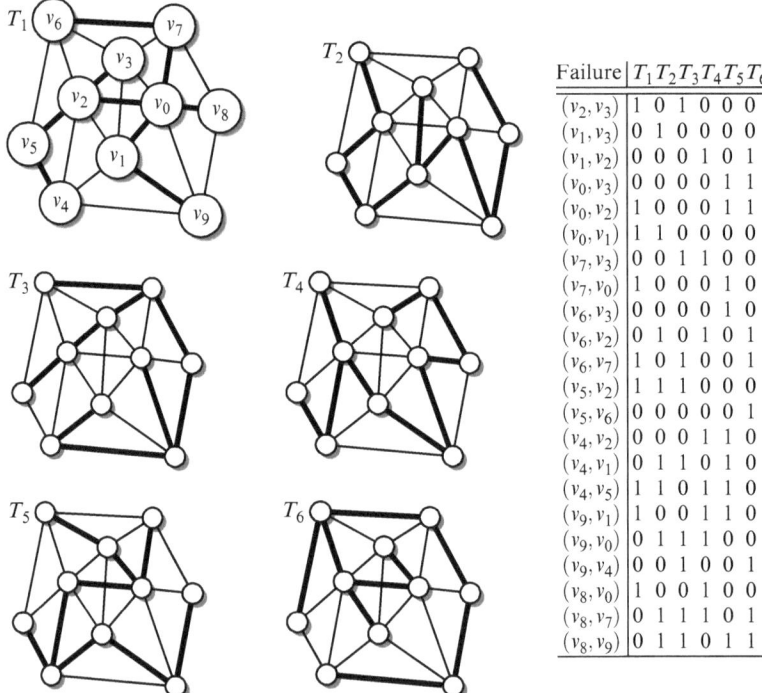

Fig. 4.2 Network-wide Local UFL (NL-UFL) via bm-trails. Unambiguous failure localization (UFL) based on bm-trails. The solution has $b = 6$ and $\|\mathscr{T}\| = 54$

4.3.1 Lower Bound for General Graphs

Definition 4.1. Let $\delta_j(\mu)$ denote 1 plus the number of links whose shortest distance from node v_j is at least μ. The shortest distance between the node and a link is defined as the length of the shortest path between the node and any of the adjacent nodes of the link. Note that $\delta_j(\mu) \geq 1$.

Theorem 4.2. Let $G = (V, E)$ be a connected network. Let $\mathscr{T} = T_1, \ldots, T_b$ be a valid bm-trail solution for NL-UFL. We have

$$\|\mathscr{T}\| \geq \sum_{j=1}^{|V|} \sum_{\mu=1}^{|V|} \frac{\mu - 1}{\mu} \left(\lceil \log_2 \delta_j(\mu - 1) \rceil - \lceil \log_2 \delta_j(\mu) \rceil \right). \quad (4.1)$$

Proof. Let the *reputation* of T_i be denoted by $r(T_i)$, which represents the number of nodes along bm-trail T_i that are aware of the on–off status of T_i. A trivial upper bound on $r(T_i)$ is $|T_i| + 1$; or formally

$$r(T_i) \leq |T_i| + 1. \quad (4.2)$$

4.3 Bounds on Bandwidth Cost

Let us define a matrix Ω with b columns and $|V|$ rows, where

$$\omega_{i,j} = \begin{cases} \frac{|T_i|}{r(T_i)} & \text{if the } i\text{-th bm-trail traverses node } v_j, \\ 0 & \text{otherwise.} \end{cases} \quad (4.3)$$

The length of T_i can be expressed as

$$\sum_{j=1}^{|V|} \omega_{i,j} = |T_i|. \quad (4.4)$$

Thus we have

$$\sum_{i=1}^{b} \sum_{j=1}^{|V|} \omega_{i,j} = \sum_{i=1}^{b} |T_i| = \|\mathcal{T}\|. \quad (4.5)$$

Let us denote the set of bm-trails passing through node v_j by $\mathcal{T}^j$. Next, we give a lower bound on $\sum_{i=1}^{b} \omega_{i,j}$ for each node v_j, which is denoted by ω_j, formally

$$\omega_j = \sum_{i=1}^{b} \omega_{i,j} = \sum_{T_i \in \mathcal{T}^j} \frac{|T_i|}{r(T_i)} \geq \sum_{T_i \in \mathcal{T}^j} \frac{|T_i|}{|T_i| + 1}. \quad (4.6)$$

Note that the cover length is the sum of ω_j for all nodes, formally

$$\|\mathcal{T}\| = \sum_{j=1}^{|V|} \omega_j. \quad (4.7)$$

Let us denote the distance between node v_j and link $e \in E$ by $\delta(v_j, e)$, which equals to the length of the shortest path between node v_j and the closest adjacent node of e plus 1. Without loss of generality, the distance can be simply measured in hops. Note that $\delta_j(k)$ denotes 1 plus the number of links whose distance from node v_j is at least k. For example, $\delta_j(1) = 1 + |E|$ as every link is at least 0 hop from the node, and $\delta_j(2) = 1 + |E| - \Delta_j$, where Δ_j is the nodal degree of j. Let n_j denote the maximum δ distance from node v_j, thus $\delta_j(n_j + 1) = 1$ and $\delta_j(n_j) > 1$. Note that such n_j always exist because there are no links with distance n. Note that $\delta_j(\mu)$ is a monotonically nonincreasing function of μ (i.e., $\delta_j(1) \geq \ldots \geq \delta_j(n_j) \geq 1$). To localize any single failure at node v_j among links whose distance from v_j is at least μ and also to identify that there is no failure among these links, there must be $\lceil \log_2 \delta_j(\mu) \rceil$ bm-trails with length at least μ.

From $\mathcal{T}^j$ we select $\lceil \log_2 \delta_j(n_j) \rceil$ bm-trails whose length is at least n_j. By Eq. (4.6) they contribute to ω_j at least

$$\frac{n_j}{n_j + 1} \lceil \log_2 \delta_j(n_j) \rceil = \frac{n_j}{n_j + 1} (\lceil \log_2 \delta_j(n_j) \rceil - \lceil \log_2 \delta_j(n_j + 1) \rceil).$$

In a similar manner, from the remaining bm-trails in $\mathscr{T}^j$ we may select as many as $\lceil \log_2 \delta_j(n_j - 1) \rceil - \lceil \log_2 \delta_j(n_j) \rceil$ bm-trails whose lengths are at least $n_j - 1$. Their contribution to ω_j is at least

$$\frac{n_j - 1}{n_j} (\lceil \log_2 \delta_j(n_j - 1) \rceil - \lceil \log_2 \delta_j(n_j) \rceil),$$

by Eq. (4.6) again. Continuing in this way for $\mu = n_j, \ldots, 1$ we obtain a collection of bm-trails in $\mathscr{T}^j$ such that the following inequality follows:

$$\omega_j \geq \sum_{\mu=1}^{n_j+1} \frac{\mu - 1}{\mu} \left(\lceil \log_2 \delta_j(\mu - 1) \rceil - \lceil \log_2 \delta_j(\mu) \rceil \right) \tag{4.8}$$

By summing up ω_j for every node according to Eq. (4.7) we get a lower bound on cover length in Eq. (4.1). Note that the $\lceil \log_2 \delta_j(x) \rceil = 0$ if $\delta_j(x) = 1$, thus we only need to consider the case where $\delta_j(x) > 1$ in the sum.

Theorem 4.2 can be well applied when the average nodal degree is small (e.g., less than 6). We show in Sect. 4.4.3.1 that Theorem 4.2 can provide a lower bound in a form $\|\mathscr{T}\| \gtrsim \xi |V| \log_2(|E| + 1)$ where ξ is somewhere in $[0.85, 0.95]$ for random networks with average nodal degree less than 6. Note that $\frac{|T_i|}{r(T_i)} \geq \frac{1}{2}$, thus we have the following corollary to Eqs. (4.6) and (4.7).

Corollary 4.1. *Let T be a valid bm-trail solution for NL-UFL. We have*

$$\|\mathscr{T}\| \geq \frac{1}{2} |V| \log_2(|E| + 1).$$

To further improve the above theorem, next we introduce a related Combinatorial Group Testing (CGT) problem. We will first consider the lower bound on this generalized version of CGT and then apply it to the NL-UFL requirement at each node, which will give us a lower bound on the cover length for general graphs. The key idea is to define a special cost function for the bm-trails at each node such that the lower bound for L-UFL at each node can be summed up to get a lower bound on the total cover length.

4.3.2 General Lower Bound for CGT

Let us consider a non-adaptive CGT problem where the goal is to find one faulty item among a set of items with group tests, where each group test is on a set of items and has two outcomes: the test contains a faulty item or not. Note that the NL-UFL problem at each node n is a special version of CGT, where the tests are the bm-trails passing through n, and the items are the links. We have two additional constraints:

4.3 Bounds on Bandwidth Cost

- The links must form a connected set in the topology, and
- Each of the links must have a nonzero code.

It is clear that a valid NL-UFL solution at node n is a valid CGT solution over links E.

Next, let us formalize the CGT problem with a cost function on each test. The cost of test T_i depends on its size according to a given cost function $\omega()$. The input of the CGT problem is a set of items denoted by $E = \{e_1, \ldots, e_m\}$ and a cost function ω, where $m = |E|$ is the number of items. The goal is to establish a set of b group tests, denoted by $T_1, \ldots, T_b$, where each group test consists of a set of items, such that a single faulty item can be unambiguously identified according to the outcomes of the group tests. It is also called *separating* test collection. Each test has a cost defined as follows:

Definition 4.2. The cost of test T_i with $t_i = |T_i|$ is $\omega(t_i)$, where function ω has the following properties:

(i) $\omega(1) = 1$, meaning that test with one element has a unit cost.
(ii) $\omega(x+1) \geq \omega(x)$ for every positive integer $1 \leq x \leq m-1$. Testing a larger group cannot decrease the cost.
(iii) $\frac{\omega(x)}{x} \geq \frac{\omega(x+1)}{x+1}$ whenever $1 \leq x < m$.

The goal is to identify the faulty item with minimum cost:

$$\text{Minimize } \Omega = \sum_{i=1}^{b} \omega(t_i) \qquad (4.9)$$

Note that much of the prior art focused on the cases with $\omega(t) = 1$, i.e., the cost of a test does not depend on the number of items, and thus the goal is to reduce the number of tests.

Theorem 4.3. *Suppose there are $m > 1$ items and assume (i)–(iii) holds for the cost function ω. Then for the cost of finding precisely one faulty item with group tests is at least*

$$\Omega \geq \min_{1 \leq x \leq \frac{m}{2}} \omega(x) \left(\log_2 x + \frac{m}{x} - 1 \right). \qquad (4.10)$$

Note that the minimum is taken over the integers x of the interval $[1, m/2]$.

Proof. Let us sort the tests by descending size, so that T_1 has the largest number of items while T_b has the least: we assume that

$$t_1 \geq t_2 \geq \cdots \geq t_b$$

where $t_i = |T_i|$ denotes the number of items in test T_i.

Also, we may assume that $t_i \leq \frac{m}{2}$ for every i. Indeed, a test set T_i with $|T_i| \geq \frac{m}{2}$ can be replaced by its complementary set $E \setminus T_i$. The resulting test collection still remains separating if the original one was separating.

We build up the $b \times m$ matrix by adding the rows one-by-one, and in each step we count the number of different columns in the matrix. Let f_i denote the number of different columns when the matrix has i rows, i.e., tests $T_1, \ldots, T_i$ are present, the others are not. For convenience we set $f_0 = 1$. Adding a row the number of different columns cannot decrease, thus $f_{i-1} \leq f_i$ for $i = 1, \ldots, b$. As we have a separating system, all the m columns will be different when the last row is added, giving that $f_b = m$.

When we add T_i, the number of different columns is at most doubled, hence $f_i \leq 2 f_{i-1}$, or

$$\log_2(f_i) - \log_2(f_{i-1}) \leq 1 \tag{4.11}$$

for $i = 1, \ldots, b$.

Similarly, by adding test T_i to the collection $T_1, \ldots, T_{i-1}$ can increase the number of different columns in the matrix by at most t_i, giving $f_i \leq f_{i-1} + t_i$, or

$$\frac{f_i - f_{i-1}}{t_i} \leq 1 \tag{4.12}$$

for $i = 1, \ldots, b$.

Now fix an integer k with $1 \leq k < b$. We have

$$\Omega = \sum_{i=1}^{b} \omega(t_i) \geq \sum_{i=1}^{k} \omega(t_i) \left(\log_2(f_i) - \log_2(f_{i-1})\right) + \sum_{i=k+1}^{b} \omega(t_i) \left(\frac{f_i - f_{i-1}}{t_i}\right). \tag{4.13}$$

We used (4.11) in the first sum, and (4.12) in the second.

The sequence $\omega(t_i)$ is nonincreasing for $i = 1, \ldots, b$ by (ii) and our numbering of the tests, hence

$$\sum_{i=1}^{k} \omega(t_i)(\log_2(f_i) - \log_2(f_{i-1})) \geq \sum_{i=1}^{k} \omega(t_k)(\log_2(f_i) - \log_2(f_{i-1})) = \omega(t_k) \log_2(f_k). \tag{4.14}$$

Similarly, the sequence $\frac{\omega(t_i)}{t_i}$ is nondecreasing because of (iii) and our numbering of the tests, giving that

$$\sum_{i=k+1}^{b} \omega(t_i) \left(\frac{f_i - f_{i-1}}{t_i}\right) \geq \sum_{i=k+1}^{b} \omega(t_{k+1}) \left(\frac{f_i - f_{i-1}}{t_{k+1}}\right) = \frac{\omega(t_{k+1})}{t_{k+1}}(m - f_k). \tag{4.15}$$

By substituting (4.14) and (4.15) into (4.13), we have

$$\Omega \geq \omega(t_k) \log_2(f_k) + \frac{\omega(t_{k+1})}{t_{k+1}}(m - f_k) \geq \omega(t_k) \left(\log_2(f_k) + \frac{m - f_k}{t_k}\right). \tag{4.16}$$

4.3 Bounds on Bandwidth Cost

This inequality is valid for any k with $1 \leq k < b$. Let us set now k to be the first index j for which $t_j \leq f_j$. Such index clearly exists and $k < b$ because $f_{b-1} \geq \frac{m}{2}$, while $t_i \leq \frac{m}{2}$ for every i. We need to consider two cases:

1. Suppose first $f_{k-1} \leq t_k$. We start from

$$\Omega \geq \omega(t_k)\left(\log_2(f_k) + \frac{m - f_k}{t_k}\right).$$

Note that $t_k \leq f_k \leq 2f_{k-1} \leq 2t_k$, hence for δ defined by $f_k = t_k + \delta$ we have $0 \leq \delta \leq t_k$. Moreover,

$$\Omega \geq \omega(t_k)\left(\log_2(t_k + \delta) + \frac{m - t_k - \delta}{t_k}\right) = \omega(t_k)\left(\log_2(t_k + \delta) - \frac{\delta}{t_k} + \frac{m}{t_k} - 1\right). \tag{4.17}$$

On the interval $0 \leq x \leq 1$ we have the inequality $x \leq \log_2(1 + x)$. We apply this for $x = \frac{\delta}{t_k}$. Note that $0 \leq \delta \leq t_k$ implies that $0 \leq x \leq 1$. We obtain the following inequality

$$\log_2(t_k + \delta) - \frac{\delta}{t_k} \geq \log_2(t_k + \delta) - \log_2\left(1 + \frac{\delta}{t_k}\right) =$$

$$= \log_2\left(\frac{t_k + \delta}{1 + \frac{\delta}{t_k}}\right) = \log_2\left(\frac{t_k + \delta}{\frac{t_k + \delta}{t_k}}\right) = \log_2(t_k). \tag{4.18}$$

Substituting (4.18) into (4.17) we get

$$\Omega \geq \omega(t_k)\left(\log_2(t_k) + \frac{m}{t_k} - 1\right).$$

2. If $t_k < f_{k-1}$, then $k > 1$ because $f_0 = 1$ by definition. Thus $f_{k-1} < t_{k-1}$ and based on (4.16) we have

$$\Omega \geq \omega(t_{k-1})\log_2(f_{k-1}) + \frac{\omega(t_k)}{t_k}(m - f_{k-1}).$$

Since $\frac{\omega(t)}{t}$ is a nonincreasing function of t, we have

$$\Omega \geq \omega(f_{k-1})\log_2(f_{k-1}) + \frac{\omega(f_{k-1})}{f_{k-1}}(m - f_{k-1}) \geq$$

$$\geq \omega(f_{k-1})\left(\log_2(f_{k-1}) + \frac{m}{f_{k-1}} - 1\right). \tag{4.19}$$

In both cases there is an integer x in the interval $[1, \frac{m}{2}]$ such that

$$\Omega \geq \omega(x) \left(\log_2(x) + \frac{m}{x} - 1\right).$$

This is because $f_{k-1} < t_{k-1} \leq \frac{m}{2}$ and $t_k \leq \frac{m}{2}$ hold. This proves the theorem.

4.3.3 Improved Lower Bound for Sparse Graphs

First let us prove two lemmas needed for the next theorem.

Lemma 4.1. *Let $m \geq 2$ fixed positive integer number. Then for $f(t) = \frac{2t}{t+1}\left(\log_2(t) + \frac{m-t}{t}\right)$ we have*

$$f(1) > f(2) > \cdots > f(\lfloor m/2 \rfloor) \geq f(m/2).$$

Proof. One can verify the lemma directly for $m = 1, \ldots, 9$. For $m \geq 10$ we have

$$f'(t) = \frac{2}{(t+1)^2}\left(\log_2 t + \frac{m-t}{t}\right) + \frac{2t}{t+1}\left(\frac{1}{t \ln 2} - \frac{m}{t^2}\right),$$

and

$$\frac{(t+1)^2}{2} f'(t) = \log_2 t + \frac{m-t}{t} + \frac{t+1}{\ln 2} - \frac{m(t+1)}{t} = \log_2 t + \frac{t+1}{\ln 2} - m - 1. \tag{4.20}$$

We have to show that

$$\log_2 t + \frac{t+1}{\ln 2} - m - 1 \leq 0 \tag{4.21}$$

on $[1, \frac{m}{2}]$. As the function of t on the left-hand side is increasing on the interval, it is enough to verify (4.21) for $t = m/2$. This follows by noting that the function $h(m) := \log_2 m - 2 + \frac{m/2+1}{\ln 2} - m$ is decreasing for $m > 8$, and $h(10) < 0$.

Lemma 4.2. *Let $n, m > 0$ be fixed real numbers. Then $\frac{2t}{n+1}\left(\log_2 t + \frac{m-t}{t}\right)$ is an increasing function of t for on $[1, \infty)$.*

Proof. Clearly it is enough to show that $g(t) = t\left(\log_2 t + \frac{m-t}{t}\right)$ is increasing. We have

$$\frac{d}{dt} g(t) = \log_2 t + \frac{m-t}{t} + t\left(\frac{1}{t \ln 2} - \frac{m}{t^2}\right) =$$

$$= \log_2 t + \frac{1}{\ln 2} - 1 > \log_2 t + 0.44 > 0$$

on $[1, \infty)$.

4.3 Bounds on Bandwidth Cost

Theorem 4.4. *The total cover length for an NL-UFL solution $\mathscr{T} = \{T_1, \ldots, T_b\}$ is at least*

$$\|\mathscr{T}\| \geq \begin{cases} \dfrac{|V| \cdot |E|}{|E|+2} \log_2 |E|, & \text{for } |V|-1 \geq \dfrac{|E|}{2}, \quad (4.22a) \\[2mm] |E| + (|V|-1) \log_2 \left(\dfrac{|V|-1}{2}\right) & \text{otherwise.} \quad (4.22b) \end{cases}$$

Proof. Let $\omega(|T_i|)$ be a cost function for bm-trail T_i as follows:

$$\omega(|T_i|) = \begin{cases} \dfrac{2|T_i|}{1+|T_i|} & \text{if } |T_i| \leq |V|-1, \quad (4.23a) \\[2mm] \dfrac{2|T_i|}{|V|} & \text{otherwise.} \quad (4.23b) \end{cases}$$

Note that as a consequence of (4.2) we have

$$\omega_{i,v} \geq \frac{\omega(|T_i|)}{2}, \text{ if } v \in T_i. \quad (4.24)$$

Next we take (4.5) to get

$$\|\mathscr{T}\| = \sum_{i=1}^{b} \sum_{v=1}^{|V|} \omega_{i,v} = \sum_{v=1}^{|V|} \left(\sum_{i=1}^{b} \omega_{i,v} \right) = \sum_{v=1}^{|V|} \left(\sum_{i|v \in T_i} \omega_{i,v} \right) \geq$$

$$\geq \sum_{v=1}^{|V|} \left(\sum_{T_i|v \in T_i} \frac{\omega(|T_i|)}{2} \right) \geq \frac{|V|\Omega}{2}, \quad (4.25)$$

where Ω is a lower bound on $\sum_{T_i|v \in T_i} \omega(|T_i|)$. The first inequality is a consequence of (4.24). Note that the function $\omega(t)$ satisfies the conditions in Definition 4.2, because $\omega(t+1) \geq \omega(t)$, $\omega(1) = 1$, and $\frac{\omega(t+1)}{t+1} \leq \frac{\omega(t)}{t}$. Although the problem with bm-trails is slightly different from finding a separating system, as in the bm-trail problem none of the items can have all zero code. However, such a constraint further restricts the problem, thus the lower bounds derived in Theorem 4.3 remain valid here as well.

Consider the case (4.22a). Then we have $t = |T_i| \leq \frac{|E|}{2} \leq |V|-1$, hence the cost function here is (4.23a). By Theorem 4.3 we have

$$\Omega \geq \min_{1 \leq t \leq \frac{|E|}{2}} \frac{2t}{1+t} \left(\log_2 t + \frac{|E|}{t} - 1 \right). \quad (4.26)$$

Here inside the min there is a decreasing function on the integer values of the interval by Lemma 4.1. It gives

$$\Omega \geq \frac{2\frac{|E|}{2}}{\frac{|E|}{2}+1}\left(\log_2\left(\frac{|E|}{2}\right)+\frac{|E|}{\frac{|E|}{2}}-1\right) =$$

$$= \frac{2|E|}{|E|+2}(\log_2|E|-1+2-1) = \frac{2|E|}{|E|+2}\log_2|E|. \quad (4.27)$$

Putting it together with (4.25) we get (4.22a).
We prove (4.22b) by applying Theorem 4.3. We obtain

$$\Omega \geq \min_{1\leq t\leq \frac{m}{2}} \omega(t)\left(\log_2 t + \frac{m}{t} - 1\right) =$$

$$= \min\left\{\min_{1\leq t\leq n-1} \frac{2t}{1+t}\left(\log_2 t + \frac{m}{t} - 1\right), \min_{n-1\leq t\leq \frac{m}{2}} \frac{2t}{n}\left(\log_2 t + \frac{m}{t} - 1\right)\right\}$$
(4.28)

where inside the first min there is a decreasing function for integer values of t by Lemma 4.1, while inside the second min there is an increasing function of t according to Lemma 4.2. The minimum is attained at $t = n - 1$. It leads to

$$\Omega \geq \frac{2(n-1)}{n}\left(\log_2(n-1) + \frac{m}{n-1} - 1\right) = \frac{2}{n}\left((n-1)\log_2(\frac{n-1}{2}) + m\right).$$
(4.29)

Putting it together with (4.25) we get (4.22b).

4.3.4 Lower Bound for Dense Graphs

Theorem 4.5 provides a lower bound on $\|\mathcal{T}\|$ as a linear function of $|E|$. With a single-link failure to be identified, any link must be traversed by at least one bm-trail. A link is called *singular* if it is traversed by exactly one bm-trail. The following two lemmas are related to singular links.

Lemma 4.3. *A bm-trail T_j of a valid bm-trail solution could traverse no more than one singular link.*

Proof. Let links e and e' be two singular links on T_j in a valid bm-trail solution. Since T_j is the only bm-trail traversing the two links, the two links cannot be distinguished when failure occurs to either one of the links. Thus the bm-trail solution is not valid, which contradicts the assumption.

4.3 Bounds on Bandwidth Cost

Lemma 4.4. *Let T_j be a bm-trail of a valid bm-trail solution. If T_j traverses a singular link, then T_j must be a spanning subgraph (i.e., connected to all the nodes) with $|T_j| \geq |V| - 1$.*

Proof. If e is a singular link of T_j, and v is a node not in T_j, then the failure of e cannot be detected by v via any bm-trail. This contradicts the fact that the bm-trail solution is valid. Since T_j is a spanning subgraph, we have $|T_j| \geq |V| - 1$.

Theorem 4.5. *Let $\mathcal{T}$ be a valid bm-trail solution for NL-UFL on a connected graph with $|E|$ links. We have*

$$\|\mathcal{T}\| \geq 2|E|\left(1 - \frac{1}{|V|}\right)$$

Proof. Let the number of singular links be denoted as σ. The cover length of a bm-trail solution can be estimated in terms of σ as follows:

$$\|\mathcal{T}\| \geq 2(|E| - \sigma) + \sigma = 2|E| - \sigma. \tag{4.30}$$

A direct consequence of Lemmas 4.3 and 4.4 is:

$$\|\mathcal{T}\| \geq \sigma(|V| - 1). \tag{4.31}$$

In case $\sigma \geq \frac{2|E|}{|V|}$, according to Eq. (4.31) we have

$$\|\mathcal{T}\| \geq (|V| - 1) \cdot \frac{2|E|}{|V|} = 2|E|\left(1 - \frac{1}{|V|}\right).$$

While in case $\sigma < \frac{2|E|}{|V|}$, according to Eq. (4.30) we have

$$\|\mathcal{T}\| \geq 2|E| - \sigma \geq 2|E|\left(1 - \frac{1}{|V|}\right).$$

Thus, for any value of σ the statement holds.

4.3.5 Line Graphs

The line graph P_n has nodes $v_1, \ldots, v_n$, and the links are (v_i, v_{i+1}) for $i = 1, \ldots, n-1$.

Theorem 4.6. *The optimal cover length for line graph P_n with n nodes and $|E| = n - 1$ links is $|E|^2$.*

Proof. For P_n the bm-trails $T_{1,2}, \ldots, T_{1,n}; T_{2,n}, \ldots T_{n-1,n}$ form a valid bm-trail solution for NL-UFL, where $T_{i,j}$ is the subpath between v_i and v_j.

If (v_i, v_{i+1}) is the faulty link, then at v_1 we recognize this correctly from the fact that $T_{1,i}$ is on, while $T_{1,i+1}$ is off. Every node between v_1 and v_i can tap these bm-trails and thus achieve L-UFL. $T_{i+1,n}$ and $T_{i,n}$ can be used similarly at the right side of the line graph, in particular at v_n.

The above bm-trail solution is optimal in the sense that if a feasible solution does not contain $T_{1,i}$ for some i (or an $T_{j,n}$ for some j) then we cannot have NL-UFL for P_n. If $T_{1,i}$ is not in the bm-trail solution, then at v_1 we cannot distinguish the failure of (v_{i-1}, v_i) from the failure of (v_i, v_{i+1}), if $i < n$. If $T_{1,n}$ is not in the m-trail solution, then v_1 cannot tell the difference between the errorless state of P_n and the failure of (v_{n-1}, v_n). Similar considerations apply to the paths $T_{j,n}$.

The cost of the collection $T_{1,2}, \ldots, T_{1,n}; T_{2,n}, \ldots T_{n-1,n}$ is

$$1+2+\cdots+n-1+1+2+\cdots n-2 = \frac{n(n-1)}{2} + \frac{(n-1)(n-2)}{2} = (n-1)^2 = |E|^2. \tag{4.32}$$

4.3.6 Stars

Let the nodes and links of star S_n be denoted by $v_c, v_1, \ldots, v_{n-1}$, and (v_c, v_i) for $i = 1, \ldots, n-1$, respectively. We note the simple facts that every set of links of S_n spans a connected subgraph; every link is adjacent to the center v_c; and a link set is adjacent to v_i, iff it contains the link (v_c, v_i). These observations imply that any set of links is a valid bm-trail, and thus the problem is simplified to a coding process for the links without considering whether a set of links can form a bm-trail. We will provide a lower and an upper bound on the cover length, which is sharp in the case $|E| = 2^{b'}$ where b' is an integer.

Lemma 4.5. *A feasible bm-trail solution on S_n has a lower bound on the total cover length:* $\|\mathscr{T}\| \geq |E|\lceil\log_2(|E|+1)\rceil$.

Proof. Any node in S_n is on at least $\lceil\log_2(|E|+1)\rceil$ bm-trails (each providing a bit of information in its ACT) to uniquely identify $|E|+1$ possible states in the network. This is according to the binary coding mechanism to perform UFL in a graph with m links. Since each node v_i is adjacent to a single-link for $i = 1, \ldots, n-1$, the number of bm-trails traversing through each link is at least $\lceil\log_2(|E|+1)\rceil$; and the total cover length is $\|\mathscr{T}\| \geq |E|\lceil\log_2(|E|+1)\rceil$ since there are $|E|$ links. Thus, the lemma is proved.

In the following a bm-trail construction is developed which exactly yields a solution with $\|\mathscr{T}\| = |E|(\lceil\log_2|E|\rceil + 1)$.

Lemma 4.6. *A bm-trail allocation construction can be found to achieve NL-UFL in S_n with cover length* $\|\mathscr{T}\| \leq |E|(\lceil\log_2|E|\rceil + 1)$.

4.3 Bounds on Bandwidth Cost

Proof. The proposed bm-trail construction is introduced as follows. Let the alarm code a_e be a bit vector of length b assigned to link e, $\forall e \in E$ such that the i-th bit is 1 if $e \in T_i$, and 0 if $e \notin T_i$. The construction is to determine how the alarm code for each link should be designed, which uniquely determines the set of bm-trails $\mathcal{T} = \{T_1, T_2, \ldots, T_b\}$.

Let us define $b' = \lceil \log_2 |E| \rceil$. The construction has alarm codes with a length $b = 2b' + 1$. We have the first b' bits uniquely code the $|E|$ links. Such a link coding process must be feasible since $2^{b'} \geq |E|$. The next b' bits will be exactly the complements of the first b' bits allocated to e. For example, if a_e has the first bit as 0, then its $(b' + 1)$-th bit should be 1, and so on. This results in the fact that the bm-trail corresponding to the i-th bit position (denoted as T_i) is the complement graph of bm-trail $T_{i+b'}$, $\forall 1 \leq i \leq b'$. Finally, the last bit of a_e is 1 for every e. With the $2b' + 1$ bm-trails, the construction is complete.

To prove the construction yields a feasible solution, our first step is to argue that every node in the star can perform L-UFL. Clearly, the number of 1's in each alarm code is $b' + 1$; in other words, exactly $b' + 1$ bm-trails are terminated at each leaf node. By further considering that each link is uniquely coded in the construction (using the first b' bits only), we can conclude that each leaf node can obtain sufficient information to perform UFL based on the terminated $b' + 1$ m-trails. Indeed, two alarm codes which agree on the $b' + 1$ bit positions which are 1 in a_e must agree everywhere because of complementation. This implies also L-UFL for the center which is traversed by all the $2b' + 1$ bm-trails.

With the construction, the number of 1's in the alarm codes of all the $|E|$ links is $|E|(b' + 1) = |E|(\lceil \log_2 |E| \rceil + 1)$, which is also the total cover length of the bm-trail solution.

Theorem 4.7. *The optimal cover length for star graph with n nodes is*

$$\|\mathcal{T}\| = |E|(1 + \lceil \log_2 |E| \rceil)$$

if $|E| = 2^{b'}$, where b' is an integer.

Proof. Based on Lemmas 4.5 and 4.6, we conclude that the construction is optimal when $|E| = 2^{b'}$ because the upper and lower bounds are equal $\|\mathcal{T}\| = |E|(\lceil \log_2(|E| + 1) \rceil) = |E|(b' + 1)$.

4.3.7 Complete Graphs

Theorem 4.8. *The minimal cover length for complete graph K_n with n nodes is*

$$\|\mathcal{T}\| = (|V| - 1)^2.$$

Proof. We prove the theorem with a construction introduced as follows. Let $S(i)$ denote the star in K_n centered at v_i. $S(v_i)$ simply consists of $n - 1$ links adjacent to v_i. Now let us arbitrarily fix a node v_w in K_n. The bm-trail solution, $\mathcal{T}$, is composed of a set of stars each centered at $\forall v_i \neq v_w$. Clearly, each star in $\mathcal{T}$ is a connected subgraph in K_n and is taken as a bm-trail. It is clear that with the construction, $\mathcal{T}$ has $n-1$ stars each covering $n-1$ links adjacent to the center of the star; thus, we have $\|\mathcal{T}\| = (n-1)^2$.

Such $\mathcal{T}$ is a valid bm-trail solution for NL-UFL in K_n. Clearly with the construction, every link must be traversed by two bm-trails except for the links adjacent to v_w, and each star is a spanning graph in K_n such that each node can access the status of all the bm-trails. Thus, the failure on any link, say (v_i, v_j), can be uniquely identified by any node in K_n due to either (1) the off status of $S(v_i)$ and $S(v_j)$ if $v_i \neq v_w$ and $v_j \neq v_w$, or (2) the off status of $S(v_i)$ or $S(v_j)$ in case $v_j = v_w$ or $v_i = v_w$, respectively. The validity of the construction is thus proved.

For optimality, by Theorem 4.5 we have

$$\|\mathcal{T}\| \geq 2m\left(1 - \frac{1}{n}\right) = n(n-1)\left(1 - \frac{1}{n}\right) =$$

$$= n(n-1) - (n-1) = (n-1)^2 = (|V|-1)^2. \quad (4.33)$$

Thus we proved the theorem.

4.3.8 Circulant Graphs

The construction in Sect. 3.2.6 is a feasible NL-UFL solution because each bm-trail spans the whole network following the unique alarm code of each link. Next we show that the proposed construction yields essentially optimal NL-UFL solutions in terms of the total cost.

Corollary 4.2. *The bm-trail construction in Theorem 3.7 is an essentially optimal NL-UFL solution.*

Proof. The average nodal degree is 4, thus $|E| = 2|V|$. The total monitoring capacity of the construction of Theorem 3.7 is $1 + |V|\lceil\log_2(2|V|+1)\rceil$. According to Theorem 4.4 the total cost is at least

$$|E| + (|V|-1)\log_2\left(\frac{|V|-1}{2}\right) = 2(|V|-1) + 2 + (|V|-1)(\log_2(2|V|-2) - 2)$$

$$= 2 + (|V|-1)(\log_2(2|V|-2)) \quad (4.34)$$

Clearly, the gap is $O(|V|)$.

4.4 The RSTA-GLS Heuristic Approach for NL-UFL

The section presents a novel heuristic algorithm to solve the NL-UFL bm-trail allocation problem in general 2-connected graphs. Inspired by the optimal solution for complete graphs in the previous section, we have observed the great suitability of using spanning subgraphs as bm-trails with bidirectional lightpaths. Thus, the proposed approach uses spanning trees as basic structures. To be specific, the proposed approach is based on two novel mechanisms referred to as *RSTA* and *GLS*, aiming to take the best advantage of flexible and cost-effective spanning tree structures, so as to overcome the topology diversity and maximize sharing of information in solving the bm-trail allocation problem for NL-UFL.

With RSTA, b randomly generated spanning trees are launched in the network. The set of b randomly generated spanning trees, denoted as $\mathscr{T}' = [T_1, T_2, \ldots, T_b]$, is used to determine the initial assignment of alarm codes for every link e (denoted as a_e) where the alarm code a_e has the j-th bit as 1 if T_j traverses through e, and 0 otherwise. Clearly, a_e is of length b bits. We claim that $\mathscr{T}'$ can be taken as a valid bm-trail solution for NL-UFL if $a_e \neq a_f, \forall e \neq f \in E$. Note that the global code uniqueness, as well as the spanning nature of each T_j, sufficiently guarantees the validity of the solution for L-UFL at each node.

Let us define a *collision* of two codes if they are identical and used by at least two links. Let the bitwise pair at the i-th position of a_e be denoted as $a_{e,[i]}$, which is the code with all identical bits as a_e except for the i-th bit. For example, 011100 is the bitwise pair for the third position of 010100.

The GLS aims to remove all possible collisions by swapping collided codes with unused ones while maintaining the spanning nature of each bm-trail. The GLS is performed iteratively upon each bit position $i = 1, \ldots, b$ one by one, where two tasks are defined in each iteration. (1) The GLS checks each bit of a link code a_e under collision, and swaps the link code with $a_{e,[j]}$ if it can resolve the code collision. (2) The code swapping in the first task will turn T_i into a subgraph with either a cycle or two isolated components. Thus GLS swaps another link code a_f on the i-th position with $a_{f,[j]}$ which is currently unused such that T_i remains a spanning tree.

4.4.1 Algorithm Description

Algorithm 3 shows the pseudo code of the proposed heuristic algorithm. With each bm-trail a spanning subgraph, the cover length is at least $\|\mathscr{T}\| \geq b(|V| - 1)$; therefore, we expect that a smaller b leads to a smaller total cover length, at the expense of smaller opportunities for successful GLS due to a smaller number of unused codes that can be taken for replacement. Two lower bounds on b are identified: the first is $b \geq \lceil \log_2(|E| + 1) \rceil$, which comes from the fact that each link must have a unique nonzero alarm code; the second is $2^{b-1} \geq |V| - 1$, which can be argued in the following lemma.

Algorithm 3: M-trail design problem for L-UFL

Input: $G = (v, E)$
begin
1.   Set $b_{ini} := \min\{\lceil \log_2(|V|-1) \rceil + 1, \lceil \log_2(|E|+1) \rceil\}$
    for $b := b_{ini}$ *to* $|V|-1$ **do**
2.   RSTA: Randomly generate b spanning trees
3.   Sort the alarm codes in descending order
4.   **for** $i := 1$ *to 500* **do**
            for *iterate through the sorted alarm codes* **do**
5.   **if** *link e and f have the same code* **then**
6.   call $GLS(G, e, i)$
                end
            end
            if *every link has a unique alarm code* **then**
7.   return **succeed**
            end
        end
    end
end

Lemma 4.7. $2^{b-1} \geq |T_j| \geq |V| - 1$ *holds for* $j = 1, \ldots, b$.

Proof. Since T_j is a spanning subgraph, we have $|T_j| \geq |V| - 1$. Further, $|T_j|$ is upper bounded by the number of codes with 1s at the bit position j, which is at most 2^{b-1}. Thus we have $2^{b-1} \geq |T_j| \geq |V| - 1$.

With the two lower bounds, the initial value of b in Step (1) is set as

$$b = \min\{\lceil \log_2(|V|-1) \rceil + 1, \lceil \log_2(|E|+1) \rceil\}.$$

In case the algorithm does not succeed with a particular b, the algorithm starts over again by increasing b, and this iteration continues until either we find a valid solution in Step (7) or $b = |V|$. With such initial setting of b, the minimum achievable cover length $||\mathscr{T}^*||$ can be expressed as:

$$||\mathscr{T}^*|| \geq (|V|-1) \min\{\lceil \log_2(|V|-1) \rceil + 1, \lceil \log_2(|E|+1) \rceil\} \quad (4.35)$$

In Step (2), b random spanning trees are generated, and each link is assigned with a code of length b where the ith bit is 1 if the link is traversed by T_i. In our implementation the method of Aldous/Broder [3,5] is adopted for this purpose. The code assignment in Step (2) may cause code collision (i.e., a code assigned to two or multiple links); and some links may not be traversed by any spanning tree (with an all-zero code "00...0"). From Steps (3) to (6), the collided codes are identified by sorting all the codes. For each link pair with a collided code, the GLS method is called in Step (6) to resolve the collision. The algorithm stops if the loop in Step (4) is executed over 500 times, to avoid infinite loops.

4.4 The RSTA-GLS Heuristic Approach for NL-UFL

Algorithm 4: Greedy link swapping (GLS)

Input: $G = (v, E)$ link e with collided code and iterator i
begin link e has code $a_e = \{a_{e,[1]}, \ldots, a_{e,[b]}\}$

6.1   **for** $i := 1$ **to** b **do**
         if $a_{e,[j]}$ *is currently unused and nonzero* **then**
6.2   **if** $a_{e,[i]} = 0$ **then**
                 After adding e into T_i the bm-trail has a cycle. Let L be the
6.3   path in bm-trail T_i between the adjacent nodes of e

             else $a_{e,[i]} = 1$
                 After erasing e in T_i the bm-trail falls into two parts. Let L be
                 the set of edges in G that can be used to reconnect the parts
             end
6.4   **for** *every link* $f \in L$ **do**
                 if $a_{f,[j]}$ *is unused and nonzero* **then**
6.5   flip the bits: $a_{e,[i]} = \neg a_{e,[i]}$ and $a_{f,[i]} = \neg a_{f,[i]}$
                     return **succeed**
                 end
             end
             if $i > 250 \wedge sparse(G) \wedge a_{e,[i]} = 0$ **then**
6.6   flip the bits: $a_{e,[i]} = \neg a_{e,[i]}$; return **succeed**
             end
         end
     end
     return **not succeed**
end

Algorithm 4 is called when a collision is found at link e, and the goal of the algorithm is to find an unused nonzero alarm code that can replace the old one and resolve the collision. In Step (6.1) we inspect each bit position on a_e to see if there is a nonzero and unused bitwise pair. If there is such $a_{e,[j]}$ (i.e., the bitwise pair of a_e at the i-th position), then we check how such swapping of the code pair would impact T_i. In case the swapping is to flip the bit at the i-th position from 0 to 1 (in Step (6.2)), it means the swapping simply adds link e to T_i. Since T_i is a spanning tree, adding e to T_i must create a loop for T_i. Thus, the algorithm tries to remove the loop by inspecting each link along the loop (denoted as L) except for e, to see if removing any link along L is possible, as shown in Step (6.4). If there is any link $f \in L$ for which $a_{f,[j]}$ is unused and nonzero, the pair of codes are swapped to remove the link from T_i such that T_i is still a spanning tree in Step (6.5).

In case the swapping is to flip the bit at the i-th bit position from 1 to 0 (in Step (6.3)), the swapping will remove e from T_i. Since T_i is a spanning tree where each node pair is at most connected by a single-link, removing e from T_i must break T_i into two isolated components. Then, the GLS needs to reconnect the two components. The links that can be used to reconnect the two isolated components except for e (denoted as L) are inspected one after the other, until any link f in L with an unused nonzero $a_{f,[j]}$ is found. If such link f is found successfully, the code pair is swapped in order to retain T_i as a spanning tree in Step (6.5).

In our implementation, in case the problem cannot be solved in 250 iterations (which happens when the networks are very sparse), cycles are also allowed. We call a network sparse, i.e., $sparse(G) := true$, iff the lower bound by Theorem 4.2 is sharper than the bound of Theorem 4.5. As shown in Step (6.6), a_e can be swapped with $a_{e,[j]}$ if a_e is nonzero and unused with 0 at the i-th position.

Note that in case the algorithm fails to find a way to resolve a code collision under a specific code length b, it breaks the loop and adds one more bit to the alarm code. This implies that the number of unused nonzero codes is doubled, which significantly helps the algorithm to resolve any possible code collision. We will show in the simulation that the proposed heuristic returns a valid NL-UFL solution in all the randomly generated topologies (over 2000).

4.4.2 An Illustrative Example

First let us illustrate the algorithm through an example using the graph in Fig. 4.3. Four random spanning trees are generated first (i.e., $b = 4$), each corresponding to a bm-trail as shown in Fig. 4.3a. Clearly, these bm-trails explicitly define an alarm code for each link as shown in Fig. 4.3a. The four bm-trails do not form a valid bm-trail solution due to the collisions $a_{(v_3,v_4)} = a_{(v_0,v_1)} = 0010$ and $a_{(v_5,v_3)} = a_{(v_3,v_1)} = 1011$.

The GLS first tries to swap $a_{(v_0,v_1)}$ with its bitwise pair of the first bit position, i.e., 1010, to remove the collision between $a_{(v_3,v_4)}$ and $a_{(v_0,v_1)}$. It is feasible since 1010 is currently unused. But with the swapping T_1 will no longer be a spanning tree due to the cycle $L_1 = (v_0, v_1, v_3, v_2)$. Thus, a link from L_1 should be removed, and the GLS does this by inspecting each link along L_1 except for the link (v_0, v_1). In our case, the first link inspected is (v_0, v_2) with an alarm code 1101. Since $a_{(v_0,v_2),[1]} = 0101$ has already been taken by link (v_5, v_4), we proceed to the next link (v_3, v_2) in L_1. Luckily, the bitwise pair $a_{(v_3,v_2),[1]} = 0111$ is currently unused by any link. Since both of the tasks in the iteration are feasible, the GLS then finalizes the swapping attempts for $a_{(v_0,v_1)}$ and $a_{(v_3,v_2)}$ with their unused bitwise code pairs: $a_{(v_0,v_1),[1]} = 1010$ and $a_{(v_3,v_2),[1]} = 0111$, respectively. As a result, T_1 is reshaped as shown in Fig. 4.3b.

For the collision between $a_{(v_5,v_3)}$ and $a_{(v_3,v_1)}$, the GLS first finds the bitwise pair of $a_{(v_5,v_3)}$ at the first bit position as 0011 that is currently unused. But by swapping $a_{(v_5,v_3)}$ as 0011, T_1 breaks into two components: $\{v_0, v_1, v_2, v_3, v_4\}$ and $\{v_5\}$, and the only way to reconnect them is to swap $a_{(v_5,v_4)} = 0101$ by its bitwise pair at the first bit position: 1101, which, unfortunately, has already been used. Therefore, there is no simple solution at the first bit, and the algorithm turns to find a solution from T_2. The bitwise pair of $a_{(v_5,v_3)} = 1011$ at the second bit position is 1111, which is unused. Further, due to the swapping, a loop $L_2 = (v_2, v_3, v_5, v_4)$ is formed on T_2 and needs to be removed. The GLS inspects each link on L_2 except (v_5, v_3), where the bitwise pair of $a_{(v_5,v_4)} = 0101$ at the second bit position,

4.4 The RSTA-GLS Heuristic Approach for NL-UFL

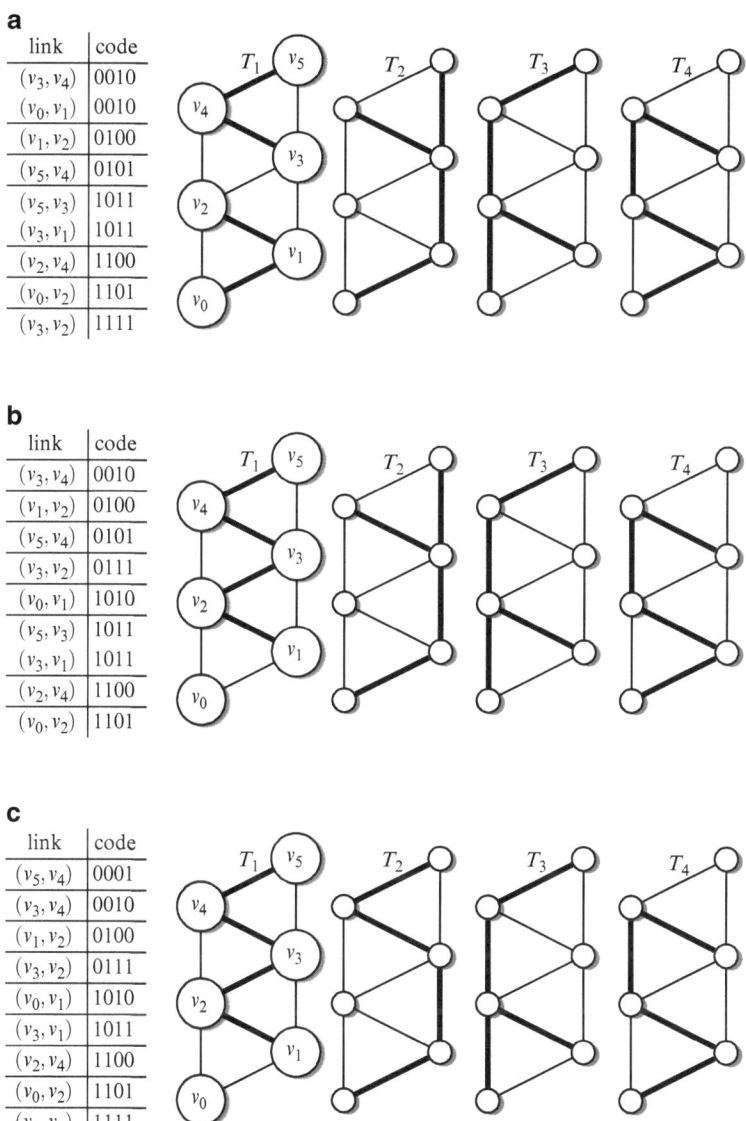

Fig. 4.3 Illustrative example of the RSTA-GLS algorithm. (**a**) Four random spanning trees. (**b**) Swapping $a_{(v_0,v_1)}$ and $a_{(v_3,v_2)}$ with their bitwise pairs at the first position. (**c**) A feasible solution by swapping $a_{(v_5,v_3)}$ and $a_{(v_5,v_4)}$ with their bitwise pairs at the second position

0001 is currently unused. Thus, the GLS concludes the feasibility of the swapping for $a_{(v_5,v_3)}$ and $a_{(v_5,v_4)}$ with their unused bitwise pairs $a_{(v_5,v_3),[2]} = 1111$ and $a_{(v_5,v_4),[2]} = 0001$, respectively.

Finally, the GLS completely solved the code collisions, and the final result is shown in Fig. 4.3c, which is a valid bm-trail solution for NL-UFL.

4.4.3 Performance Verification of RSTA-GLS

Simulations on thousands of randomly generated planar 2-connected backbone network topologies via LEMON [8] were conducted. With LEMON random graph generator, nodes are firstly allocated into an area of unit square with a uniform distribution, and links with small physical length are added to keep the graph planar if possible, and to keep the facets of the planar graph of equal size (see examples in Fig. 1.1j–l). In such a way thousands of random networks were generated and classified according to their size in number of edges or nodes and nodal degrees. Finally a series of networks are selected and stored, respectively, according to their (1) number of nodes, (2) number of links, and (3) nodal degrees. The performance metrics of interest are the total cover length of the solution and the running time. In addition to the evaluation of the proposed heuristic with the derived upper bounds and optimal solutions, the impact of topology diversity to the NL-UFL bm-trail problem will be investigated. All three metrics are examined with respect to different network sizes (i.e., the number of nodes or links) and topology densities (i.e., nodal degree), which will be presented in the following three subsections.

4.4.3.1 Performance Comparison

We first examine the proposed scheme by comparing it with the bounds derived as shown in Fig. 4.4a, b in terms of total cover length (i.e., $\|\mathcal{T}\|$). Figure 4.4a shows the cover length when the average nodal degree of the topologies with 50 nodes is increased from 2.5 to 49, where three curves by the proposed *RSTA-GLS*, Theorem 4.4, Theorem 4.2, and Theorem 4.5, are plotted, respectively. It is shown that *RSTA-GLS* yields an average gap of less than 9.7 % to the best bound of the two, which is given by Theorem 4.4 when the topology is sparse while by Theorem 4.5 in denser ones. Figure 4.4b provides a closer look at the range of nodal degree [2.5, 7] in Fig. 4.4a, where the bounds given by Theorem 4.4, Theorem 4.2, and Eq. (12) are also plotted. Note that Eq. 4.35 is a bound that intrinsically exists due to the initial assignment of b. To summarize, the proposed *RSTA-GLS* can yield solutions very close to the lower bounds that we derived, and the two bounds in Theorem 4.4 and Theorem 4.5 seem to be reasonable for sparse and dense topologies, respectively. Further, since the bounds derived for general graphs do not consider sufficient

4.4 The RSTA-GLS Heuristic Approach for NL-UFL

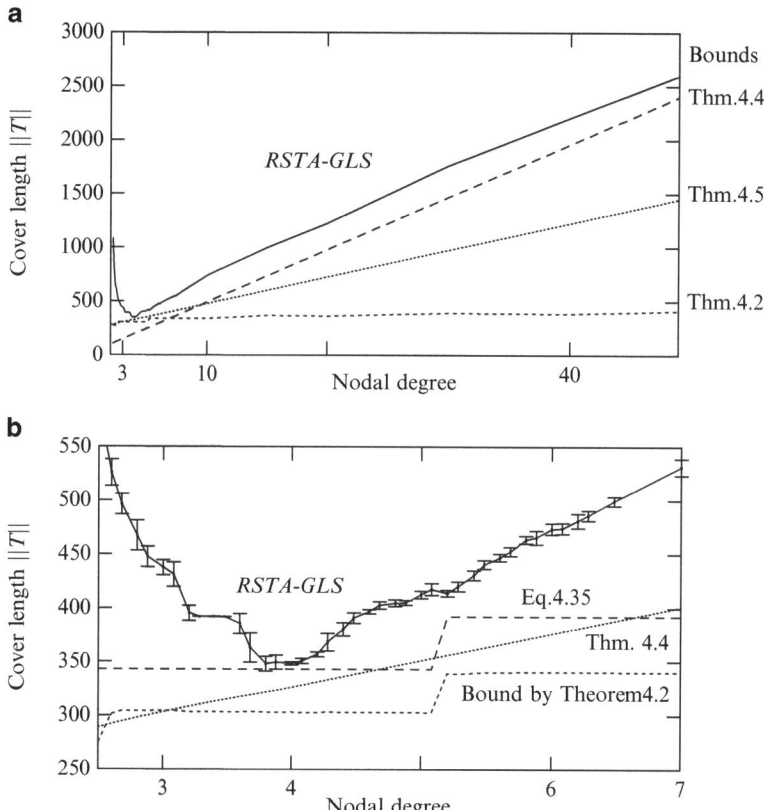

Fig. 4.4 The cover length versus average nodal degree (900 randomly generated topologies with 50 nodes). (**a**) Networks with 50 nodes and different number of links. (**b**) Plot with nodal degree from 2.5 to 7

topological features, both bounds are not effective when the network nodal degrees are quite low (e.g., less than 3). Nonetheless, the minimal gap is observed when the nodal degree ≈4.

Three schemes were implemented and compared in the simulation. *RSTA-GLS* (denoted by × on the charts) refers to the proposed heuristic. CA_S (denoted by + on the charts) corresponds to the method that each trail is allocated one after the other to distinguish each pair of links using a scheme based on Dijkstra's algorithm, such that L-UFL can be achieved at an arbitrarily chosen monitoring location. Such a method is generic and has been considered in a number of previously reported studies [1,4,16]. Rather than just for L-UFL at a single ML with CA_S, CA_N (denoted by △ on the charts) achieves NL-UFL by performing CA_S sequentially at each node. In our implementation loopback switching is allowed, thus CA_S and CA_N algorithms are modified such that an L-UFL solution for any node can always be found in a connected graph according to Theorem 4.1.

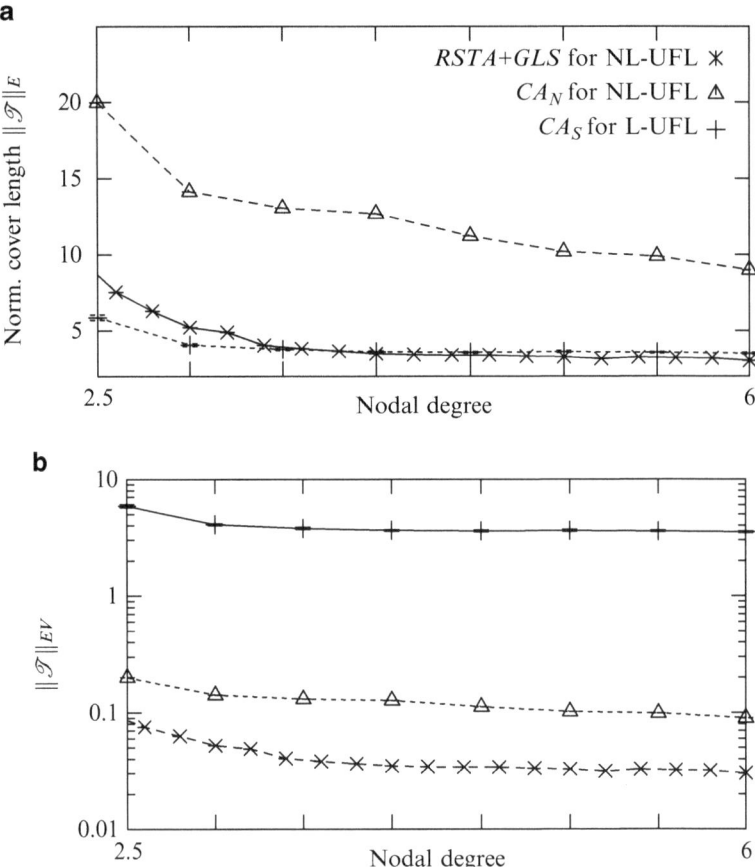

Fig. 4.5 The normalized cover length by *RSTA-GLS* (*times symbol*) and CA_S (*plus symbol*) and CA_N (*open triangle*) of 1560 randomly generated topologies with 100 nodes

The comparison results are shown in Figs. 4.5, 4.6, and 4.7. We normalize the cover length over the number of links and both the numbers of links and nodes, denoted as $\|\mathcal{T}\|_E$ and $\|\mathcal{T}\|_{Ev}$, respectively. $\|\mathcal{T}\|_E$ is a measure on the average number of WLs per edge, while $\|\mathcal{T}\|_{Ev}$ is on the average number of WLs per edge per monitoring location, respectively. Clearly we have $\|\mathcal{T}\|_E = \frac{\|\mathcal{T}\|}{|E|}$ and $\|\mathcal{T}\|_{Ev} = \frac{\|\mathcal{T}\|}{|V|\cdot|E|}$.

In Fig. 4.5a, b, we observed that the proposed *RSTA-GLS* consumes much smaller $\|\mathcal{T}\|_E$ and $\|\mathcal{T}\|_{Ev}$, respectively, than that by CA_N when the number of network nodes is 100 and the average nodal degree is increased from 2.5 to 6. It is clearly demonstrated that *RSTA-GLS*, which enables all the 100 nodes to individually serve as L-UFL monitoring locations, can achieve similar cover lengths to that by CA_S where only a single L-UFL ML is supported. This demonstrates the effect of on–off

4.4 The RSTA-GLS Heuristic Approach for NL-UFL

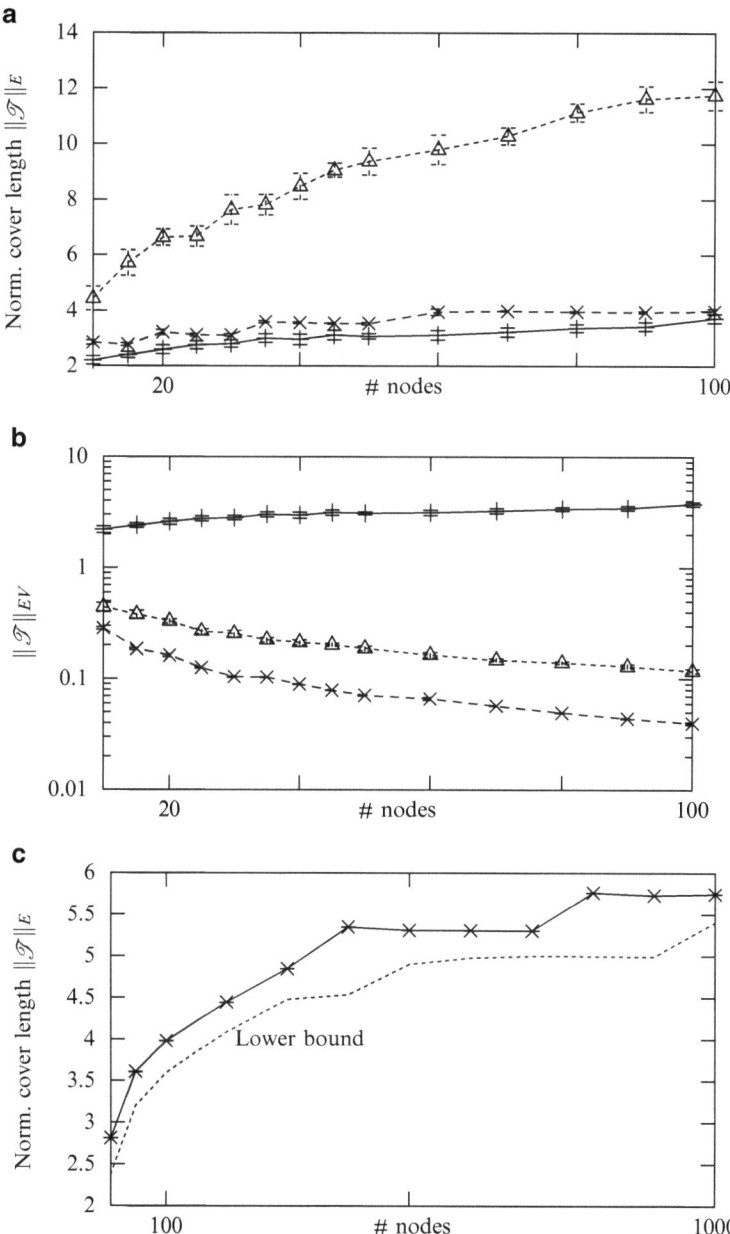

Fig. 4.6 The normalized cover length by *RSTA-GLS* (*times symbol*) and CA_S (*plus symbol*) and CA_N (*open triangle*). (**a**) and (**b**) 2400 randomly generated topologies with nodal degree ≈ 4. (**c**) 240 randomly generated topologies with nodal degree ≈ 4

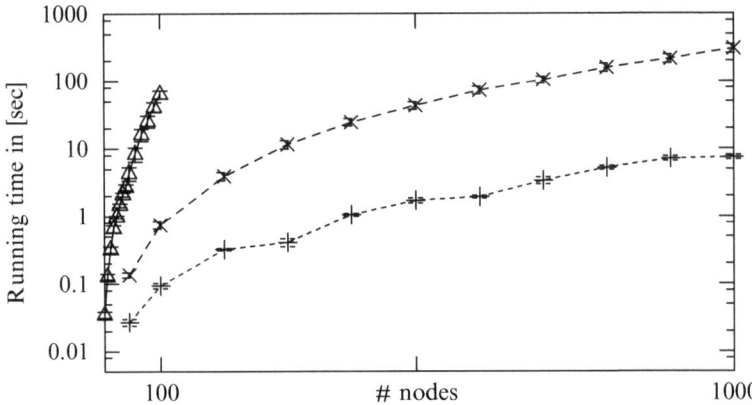

Fig. 4.7 The normalized running time by *RSTA-GLS* (*times symbol*) and CA_S (*plus symbol*) and CA_N (*open triangle*)

status sharing among the on-trail nodes of a common bm-trail, such that the increase of the number of MLs would lead to little increase in the total cover length. They also show that $\|\mathscr{T}\|_E$ and $\|\mathscr{T}\|_{Ev}$ decrease when the nodal degree is increased (or as the number of links is increased) due to the better connectivity which yields larger design space for allocating the bm-trails. Figure 4.6a, b plot $\|\mathscr{T}\|_E$ and $\|\mathscr{T}\|_{Ev}$ against the number of nodes given the nodal degree ≈ 4, which yield similar comparison results as in Fig. 4.5a, b. Figure 4.6c plots the solutions by *RSTA-GLS* and the best lower bound derived in Sect. 4.3.1. One can verify that the normalized cover length has a logarithmic relation with the increasing number of nodes (from 100 to 1,000) when the average nodal degree is kept as a constant (i.e., ≈ 4).

Finally, the computation efficiency of the proposed *RSTA-GLS* is examined, which is shown in Fig. 4.7. We can clearly see that our scheme yields significantly better computation efficiency than CA_N. Note that CA_S takes shorter computation time since only a single ML is considered, compared with the case where *RSTA-GLS* enables all the nodes as MLs each being able to perform L-UFL. With *RSTA-GLS*, the NL-UFL bm-trail problem on a 1000-node network can be solved within 3 min.

Table 4.1 provides results of *RSTA-GLS* and CA_N on some well-known network topologies taken from [12]. The number of nodes, links and the diameter in hops of every topology graph is also shown in the first three columns of the table. The results are similar to that on randomly generated graphs as in Fig. 4.4 regarding the number of required bm-trails (i.e., $\|\mathscr{T}\|$), average WLs per link (i.e., $\|\mathscr{T}\|_E$), and running time (in seconds) under various network sizes and topology densities.

To summarize the comparison results above, the proposed *RSTA-GLS* is witnessed to achieve desired computation efficiency in handling large networks, and its feasibility in the operation of future all-optical backbone is proved in terms of resource consumption for achieving NL-UFL using bidirectional m-trails. For example, an optical network with 288 links and 100 nodes takes averagely 3.4 WLs

4.4 The RSTA-GLS Heuristic Approach for NL-UFL

Table 4.1 Results by the proposed *RSTA-GLS* and CA_N on some well-known networks

Network [12]	Graph				Theorem		$\|\mathcal{T}\|$			$\|\mathcal{T}\|_E$		#m-Trails		Time (s)	
	$\|V\|$	$\|E\|$	dia-met	4.2	4.4	4.5	RSTA+GLS	CA_N	RSTA+GLS	CA_N	RSTA+GLS	CA_N	RSTA+GLS	CA_N
Pan-Europe	16	22	6	79	65.6	82	107	241	2.43	5.47	7	24	0.39	0.42
German	17	26	6	84	74	97	128	368	2.46	7.07	8	33	0.51	0.65
ARPA	21	25	7	108	91.4	95	140	354	2.80	7.08	6	33	0.36	0.64
European	22	45	5	127	116	171	231	617	2.56	6.85	11	53	2.10	4.59
USA	26	42	8	155	133	161	229	667	2.72	7.94	9	53	2.16	3.99
Nobel EU	28	41	8	168	142	158	248	689	3.02	8.40	7	50	0.87	3.38
Italian	33	56	9	200	184	217	329	1113	2.93	9.93	10	66	4.83	11.1
Cost 266	37	57	8	225	207	221	343	918	3.00	8.05	8	55	2.04	14.4
North Amer.	39	61	10	241	222	237	378	1020	3.09	8.36	8	50	2.31	15.5
NSFNET	79	108	16	579	520	426	760	2634	3.51	12.2	9	100	6.05	227

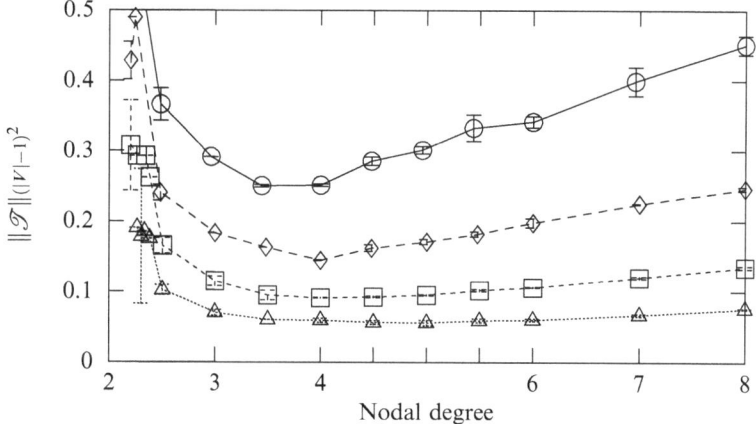

Fig. 4.8 Plot of $\|\mathcal{T}\|_{(|V|-1)^2}$ with $n = 25$ (*open circle*), 50 (*open diamond*), 100 (*open square*), 200 (*open triangle*) (1,280 randomly generated topologies)

along each link, and each ML only consumes approximately 0.034 WLs per link for achieving NL-UFL. Thus, in the case that each fiber has 100 wavelength channels and 10 fibers bundled in a single conduit, the system needs to spend only 0.68 % of the total capacity for achieving NL-UFL of any single conduit.

4.4.3.2 The Impact of Topology Diversity

Figure 4.8 shows the performance of the proposed *RSTA-GLS* on topologies with $|V| = 25, 50, 100, 200$ as the increase of the average nodal degree of the topologies. Instead of plotting $\|\mathcal{T}\|$ for each case, we normalize the results on topologies of n nodes by $(|V|-1)^2$ (which is denoted as $\|\mathscr{T}\|_{(|V|-1)^2}$). Note that a line graph P_n and a complete graph K_n represent the least and most densely meshed topology with n nodes, respectively. Both topologies have an optimal solution $\|\mathscr{T}\| = (|V|-1)^2$ as derived in Sect. 4.3.7, which is supposed to be the largest possible $\|\mathscr{T}\|$ for any topology with n nodes. We expect that all other topologies should yield solutions less than $(|V|-1)^2$. This is attested in the figure, where the curve of $\|\mathscr{T}\|_{(|V|-1)^2}$ is observed as a "V" shape for each specific n value. It also shows that $\|\mathscr{T}\|_{(|V|-1)^2}$ is minimal when the average nodal degree of the topologies is in the range of $[3.5, 4.5]$ for all the values of $|V|$ investigated. This demonstrates that the proposed scheme can be the most suitable to work for optical networks with moderate connectivity.

4.5 Summary

In this chapter, we discussed distributed failure localization, i.e., the scenario where a node can individually perform UFL. This approach eliminates control plane signaling from the failure localization phase, where the alarms are locally available at the (L-UFL capable) node, which can make the failure localization decision. However, the other nodes of the network still should be notified in the restoration process of a failure in order to perform the appropriate recovery action to recover the disrupted connections traversing that node. The NL-UFL framework, where each node is L-UFL capable, is already a good candidate as the foundation of an all-optical signaling-free framework, if it is combined with a protection approach accordingly. Incorporating distributed failure localization and state-of-the-art protection approaches will be discussed in Part III of this book.

References

1. Ahuja S, Ramasubramanian S, Krunz M (2009) Single link failure detection in all-optical networks using monitoring cycles and paths. IEEE/ACM Trans Netw 17(4):1080–1093
2. Ahuja S, Ramasubramanian S, Krunz M (2011) SRLG failure localization in optical networks. IEEE/ACM Trans Netw 19(4):989–999
3. Aldous D (1990) The random walk construction of uniform spanning trees and uniform labelled trees. SIAM J Discrete Math 3(4):450–465
4. Assi C, Ye Y, Shami A, Dixit S, Ali M (2002) A hybrid distributed fault-management protocol for combating single-fiber failures in mesh based DWDM optical networks. In: Proceedings of the IEEE GLOBECOM, pp 2676–2680
5. Broder A (1989) Generating random spanning trees. In: Annual symposium on foundations of computer science. IEEE Computer Society, Research Triangle Park, North-Carolina, USA, pp 442–447
6. Harvey N, Patrascu M, Wen Y, Yekhanin S, Chan V (2007) Non-adaptive fault diagnosis for all-optical networks via combinatorial group testing on graphs. In: Proceedings of the IEEE INFOCOM, pp 697–705

References

7. Lee K, Modiano E (2009) Cross-layer survivability in wdm-based networks. In: Proceedings of the IEEE INFOCOM, pp 1017–1025
8. LEMON (2010) a C++ Library for Efficient Modeling and Optimization in Networks. http://lemon.cs.elte.hu
9. Mao M, Yeung K (2010) Super monitor design for fast link failure localization in all-optical networks. In: Proceedings of the IEEE ICC
10. Médard M, Barry R, Finn S, He W, Lumetta S (2002) Generalized loop-back recovery in optical mesh networks. IEEE/ACM Trans Netw 10(1):164
11. Ogino N, Nakamura H (2011) All-optical monitoring path computation based on lower bounds of required number of paths. In: Proceedings of the IEEE ICC
12. Orlowski S, Pióro M, Tomaszewski A, Wessäly R (2007) SNDlib 1.0-survivable network design library. In: Proceedings of the international network optimization conference (INOC)
13. Tapolcai J (2014) Reliable telecommunication networks. DSc Dissertation, Hungarian Academy of Sciences. http://real-d.mtak.hu/573/7/TapolcaiJanos_doktori_mu.pdf
14. Tapolcai J, Rónyai L, Ho PH (2010) Optimal solutions for single fault localization in two dimensional lattice networks. In: Proceedings of the IEEE INFOCOM mini-symposium, San Diego
15. Tapolcai J, Ho PH, Rónyai L, Wu B (2012) Network-wide local unambiguous failure localization (NWL-UFL) via monitoring trails. IEEE/ACM Trans Netw 20(6):1762–1773
16. Zeng H, Huang C, Vukovic A (2006) A novel fault detection and localization scheme for mesh all-optical networks based on monitoring-cycles. Photonic Commun Netw 11(3):277–286
17. Zhou D, Subramaniam S (2000) Survivability in optical networks. IEEE Netw 14(6):16–23

Part III
An All-Optical Restoration Framework with M-Trails

With NL-UFL, every node can localize an SRLG failure in an all-optical fashion. It opens a new design dimension that a node can use the failure localization information to instantly react to the failure event in the subsequent failure restoration process without waiting for further signaling. In specific, nodes along a P-LP, once localized a failure event that interrupted the corresponding W-LP, can instantly and automatically perform the predefined switch fabric reconfiguration to form a physical lightpath for the failure restoration. Thus, the failure event can be restored completely in the optical domain without any aid from the electronic signaling, which forms an all-optical restoration framework for achieving ultra-fast failure restoration.

The chapters of Part III present and investigate the all-optical restoration framework under both static and dynamic traffic scenarios. By jointly allocating m-trails, W-LPs, and P-LPs, we will show that the P-LPs and m-trails can reuse a common WL, and the all-optical restoration framework based on a failure dependent protection (FDP) scheme can significantly outperform its counterparts. It has a great potential to achieve an ultra-fast restoration speed like in a ring network while enjoying near optimal capacity efficiency like in a mesh network.

Chapter 5
Framework Introduction

Abstract This chapter discusses an interesting application of NL-UFL for supporting all-optical and signaling-free restoration under failure dependent protection (FDP). We demonstrate a new failure restoration framework that enables all-optical fault management and device configuration in the presence of NL-UFL m-trail deployment, such that the FDP restoration process can be implemented without relying on any control plane signaling. It is also shown that the reuse of spare capacity by P-LPs for the m-trails can minimize the additionally required resources (in WLs) for m-trails, which is referred to as *monitoring resource hidden*.

5.1 Introduction

As we have discussed in Part I of this book, a general approach to increase availability of each connection is to pre-plan one or multiple protection lightpaths (P-LPs) for each working lightpath (W-LP). Fortunately, network failures are rare events, thus it is a widely accepted strategy to share the allocated spare capacity among multiple P-LP(s) that are assumed not to be activated at the same time, referred to as *shared protection*. On the other hand, a shared protection scheme relies on a suite of real-time mechanisms to restore the failed W-LPs, including failure localization, failure notification, failure correlation, and device configuration (Sect. 1.4). Remember that two objectives are widely considered in the design of a protection scheme in optical networks, namely resource efficiency and fault management complexity. The former concerns the amount of consumed *spare capacity*, which is the capacity in terms of *wavelength channels/links (WLs)* reserved but not necessarily configured for the P-LP(s); while the latter is measured as *restoration time*, which equals to the duration from the instant that the traffic is unexpectedly interrupted to that the traffic is completely restored.

In the spectrum of different flavors of shared protection schemes, there exists a compromise between capacity efficiency and restoration time, in which Failure Dependent Protection (FDP) and p-Cycle are the two extremes (Sect. 2.3). With FDP, each connection is assigned with multiple P-LPs, and one is activated for the restoration purpose according to the identified failure event. With stub release, FDP has long been recognized as having the optimal capacity efficiency and widely taken as the performance benchmark in the design of shared protection schemes.

However, the FDP restoration process is subject to the highest control and signaling complexity that possibly yields the longest restoration time, mostly because the switching node has to precisely localize the failure event for real-time selection of a P-LP.

It was open whether there is an efficient approach that can facilitate a general protection scheme to achieve an all-optical and signaling-free restoration process like p-Cycle without sacrificing the capacity efficiency of FDP. In this chapter we intend to investigate the framework of fast restoration in all-optical networks, aiming to enable any shared protection scheme to achieve true all-optical restoration without any aid from the upper layer protocol. The proposed new framework incorporates the conventional restoration process with state-of-the-art failure localization techniques, namely *Network wide Local Unambiguous Failure Localization (NL-UFL)* described in Sect. 4.2.4, which takes advantage of a set of intelligently deployed supervisory lightpaths (S-LPs), called *monitoring trails* (m-trails). The on–off status of the m-trails can be converted into network-wide failure status locally available to each node, such that each node can autonomously respond to any predefined failure event and collaboratively work for the recovery of the interrupted W-LPs. The chapter will detail how the restoration process can be realized, and will compare the proposed approach with a couple of p-Cycle-based schemes via a case study in terms of restoration time, computation complexity, and capacity efficiency.

5.2 Signaling-Free Restoration Framework

The proposed framework is characterized by an integration of the conventional restoration process with a state-of-the-art all-optical failure localization technique, NL-UFL. In a nutshell, NL-UFL creates an all-optical fault management system by taking all the nodes as independent monitoring locations. Without relying on any multi-hop signaling protocol, each node can obtain the on–off status of the traversing m-trail(s) by tapping the optical signals along the m-trail(s), which further facilitate the formation of a valid alarm code that uniquely maps to the failure event. Thus, NL-UFL leads to a truly all-optical and signaling-free failure localization system and is taken as a key functional block in the proposed framework.

Recall the restoration time analysis in Sect. 2.1 for the conventional GMPLS-based recovery where the restoration time is mainly due to failure localization (t_l), failure notification (t_n), failure correlation (t_c), and device configuration (t_d). On the other hand under the proposed framework, we show that the four terms can be significantly reduced or removed, and the resultant restoration process can be implemented completely in the optical domain without the aid of any network layer control protocol.

Firstly, since the switching node of an interrupted W-LP is aware of the failed SRLG via the on–off status of the traversing m-trails, the time for failure localization is simply the propagation delay of the m-trails interrupted by the failure, while the time for failure notification is completely removed. Since an optical flow is

5.2 Signaling-Free Restoration Framework

deterministic in terms of its propagation speed, the time for failure correlation can be the minimum. Further, since all the intermediate nodes of the P-LPs corresponding to the failure event can localize the failure event thanks to NL-UFL, they can start configuring their switch fabrics based on the collected alarm code without waiting for P-LP setup request from any other network entity. Thus the W-LP setup latency can be minimized as well. This leads to a completely all-optical and deterministic restoration process.

To be specific, the fault management latency under the proposed restoration process can be modeled as:

$$t_r = t_{nc} + t_d,$$

where $t_{nc} = t_n + t_c$ and t_d is the time for failure notification/correlation, and the time for device configuration for the P-LP setup, respectively. Different from that in the conventional restoration process, t_{nc} is determined solely by the light propagation delay of the m-trails. On the other hand, t_d should be close to the time for OXC configuration at a single node, since all the intermediate nodes of the P-LP can configure their OXCs almost in parallel.

Following up Fig. 2.1, Fig. 5.1 is an illustration for the proposed restoration process. Let there be two m-trails $T_1 : v_3 - v_2 - v_1 - v_4 - v_5 - v_3$ and $T_2 : v_2 - v_1 - v_4 - v_5 - v_3$ which help nodes v_1, v_2, v_3, v_4, and v_5 to localize the failure at link $v_2 - v_3$. With this, node v_1 can identify the failure at $v_2 - v_3$ by taking time $t_{nc}^{v_2 - v_1}$ right after the failure of $v_2 - v_3$, and node v_1 looks up its *restoration table* (i.e., extended ACT with switching information, see Sect. 5.2.1 for definition) and immediately switches over the traffic by taking another latency of t_d. Meanwhile, nodes v_4, v_5, and v_3 can also identify the failure by taking propagation time

$$t_{nc}^{v_2 - v_1 - v_4}, t_{nc}^{v_2 - v_1 - v_4 - v_5}, \text{ and } t_{nc}^{v_2 - v_1 - v_4 - v_5 - v_3},$$

respectively, and they look up their restoration tables and immediately configure their OXCs to realize the P-LP which takes t_d. Thus, the total consumed time is the longest of the propagation times plus a single device configuration time.

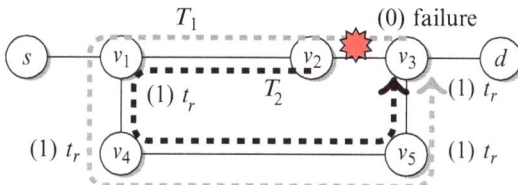

Fig. 5.1 The proposed restoration process

5.2.1 An Example on the Restoration Process

With the help of S-LPs, the proposed failure restoration process can be performed locally in the optical domain without relying on any multi-hop signaling. To achieve this, each node should maintain a *restoration table (RT)* extended from its local ACT as shown in Fig. 5.2. Compared to ACT, an RT has one more column which keeps the mapping from every localized failure state (i.e., the obtained alarm code) to the corresponding reconfiguration on the OXC. The nodes along a P-LP, including the switching node, intermediate nodes, and the merger node, can start configuring their OXCs to form the required cross-connect by looking up their RTs once a valid alarm code is obtained. With this, the P-LP can be formed right after the identification of the failure event without any further path setup mechanism. Note that the RT of a node should be updated when any connection whose P-LP traverses through the node is being set up or torn down.

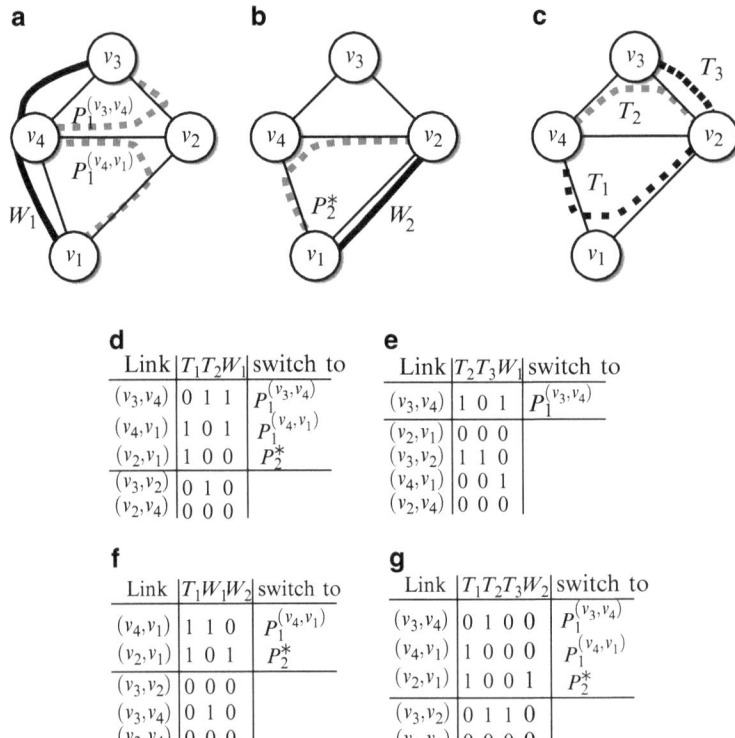

Fig. 5.2 An illustrative example for the proposed restoration process with the corresponding restoration tables at each node. (**a**) Connection 1, (**b**) Connection 2, (**c**) M-trails T_1, T_2, T_3, (**d**) RT at v_4, (**e**) RT at v_3, (**f**) RT at v_1, (**g**) RT at v_2

5.3 The Spare Capacity Allocation Problem

Figure 5.2 shows an example in a topology with four nodes and two W-LPs denoted as W_1 and W_2, each being provisioned with two physical lightpaths on the same route in opposite directions. W_1 is protected by two P-LPs, namely $P_1^{(v_3,v_4)}$ for link failure (v_3, v_4), and $P_1^{(v_4,v_1)}$ for (v_4, v_1) as in Fig. 5.2a, while W_2 is protected by a single P-LP denoted as P_2^* as in Fig. 5.2b. To ensure signaling-free restoration for W_1 and W_2, v_1 should be able to unambiguously identify the failure of (v_4, v_1) and (v_2, v_1), such that W_1 (or W_2) can be switched over to $P_1^{(v_4,v_1)}$ (or P_2^*) when (v_4, v_1) (or (v_2, v_1)) fails. Thus, three m-trails T_1, T_2, and T_3 are deployed as shown in Fig. 5.2c in order to unambiguously identify the failures for the different switching configurations.

Note that, in the example in Fig. 5.2 not all links have unique alarm codes, only those for which the given node have to respond in the restoration process. In order to precisely define which failures require UFL at a node, Chap. 6 provides more details, while the joint allocation of P-LPs and S-LPs is discussed in Chap. 7. In the rest of Chap. 5 we lay down the theoretical background of these restoration frameworks by inspecting the NL-UFL scenario, i.e., when each node has to respond to all SRLG failures in the network. These conditions can be incorporated in the design of S-LPs and P-LPs for specific failures.

5.3 The Spare Capacity Allocation Problem

In this section we define the *Spare Capacity Allocation (SCA)* problem, the interested reader can refer to [2–4,6–10,12,13,15,16,20,21,24] for more description on the heuristic algorithms solving the SCA problem.

5.3.1 The FDP-SCA Problem Formulation

A key issue to achieve the proposed FDP restoration framework is an approach to determine how m-trails and P-LPs under FDP are allocated, when we are provided with the network topology and a set of W-LPs. The spare capacity problem is different from any previously reported problems since it considers the allocation of both monitoring resources (for m-trails) and restoration resources (for P-LPs). The input of the problem is as follows.

1. The *network topology*, which is represented by an undirected graph $G = (V, E)$. For the sake of simplicity we assume infinite capacity on each link.
2. The *set of SRLGs*, which is denoted by $\mathscr{Z}$. Each SRLG $z \in \mathscr{Z}$ contains one, or at most two and adjacent links.
3. A set of W-LPs denoted as $\mathscr{W} = W_1, W_2, \ldots, W_k$, where k is the total number of W-LPs. Each W_j is a path in G between nodes s_j and d_j for $j = 1, \ldots, k$.

The SCA problem is to minimize the total spare capacity (denoted by $\mathcal{M}$) while achieving: (i) a feasible solution for NL-UFL m-trail allocation under multi-link SRLGs, and (ii) restoration capacity allocation for FDP. Formally

$$\mathcal{M} = \sum_{e \in E} u_e, \qquad (5.1)$$

where u_e denotes the spare capacity along link $e \in E$ in terms of the number of wavelength links (WLs).

An important feature of the proposed approach is that the spare capacity taken by the m-trails can be reused by any P-LP. This is a reasonable assumption since besides the working traffic either P-LPs or m-trails are launched. Once there is a failure, all monitoring resources for the m-trails can be released and reused by the P-LPs, thanks to NL-UFL which enables all the nodes to instantly react upon the identified failure. With this, we have:

$$u_e = \max\{m_e, p_e\},$$

where m_e denotes the required monitoring resources for the m-trails as

$$m_e = \sum_{1 \leq j \leq b} I^T_{j,e} \qquad (5.2)$$

where $I^T_{j,e}$ is the trail-link indicator (a.k.a global ACT) which is 1 if the jth m-trail passing through link e, and 0 otherwise. And p_e denotes the restoration capacity for FDP on link $e \in E$ that is formalized as follows.

5.3.1.1 The Restoration Capacity

Let $\mathcal{Z}_j \subseteq \mathcal{Z}$ be a subset of SRLGs traversed by W_j. For each W_j, $1 \leq j \leq k$, a set of P-LPs denoted as P^z_j, $\forall z \in \mathcal{Z}_j$, is determined such that they are simple paths with source s_j and destination d_j while disjoint from each z. Let $\mathcal{W}^z \subseteq \mathcal{W}$ be a subset of W-LPs, which are the W-LPs possibly interrupted by failure z. The amount of reserved working capacity along link e is denoted as q_e, $\forall e \in E$; formally $q_e = \sum_{1 \leq j \leq k} I^W_{j,e}$, where $I^W_{j,e}$ is a working path-link indicator which is 1 if W_j passes through link e, and 0 otherwise. Let $I^{P,z}_{j,e}$ be a protection path-link indicator which is 1 if P^z_j traverses e, and 0 otherwise. Let the restoration capacity along link e in the failure event of z be denoted by p^z_e. We have the following relation:

$$p^z_e = \sum_{1 \leq j \leq k} I^{P,z}_{j,e} - \sum_{W_j \in \mathcal{W}^z} I^W_{j,e}$$

where the term $\sum_{W_j \in \mathcal{W}^z} I^W_{j,e}$ stands for the free capacity gained by *stub release* at link e after the failure event at z which interrupted all $W_j \in \mathcal{W}^z$.

5.4 The Monitoring Resource Hidden Property

The condition for p_e to restore every affected W-LP $W_j \in \mathscr{W}^z$ by failure z is:

$$p_e \geq \max_{z \in \mathscr{Z}_j} p_e^z.$$

In general, the monitoring resource consumption is determined only by the topology and SRLGs considered in the problem regardless of the traffic, which serves as a constant expense; on the other hand, the amount of restoration resources heavily depends on the amount of working traffic. Thus, when the amount of working capacity increases, more monitoring resources will be reused due to the increased restoration resource consumption, and gradually all the monitoring resources are reused. We define the monitoring resources that are not shared by any P-LP as *monitoring overhead*, denoted by

$$r_e = \max\{0, m_e - p_e\}.$$

The monitoring resources are expected to be completely hidden by P-LPs, provided that we have sufficient working capacity in the network, resulting zero monitoring overhead. This is also referred to as the *monitoring resource hidden property* where all WLs taken by m-trails are also reserved by P-LPs.

5.3.2 FDP Restoration Capacity Allocation

As discussed in Sect. 2.3.5, the restoration capacity allocation problem for FDP has been extensively investigated in the past decades, which is formally called *Spare Capacity Placement/Allocation for Path Restoration with stub release* [5, 8, 10, 14, 24]. Since we have positioned our study on a signaling-free FDP-based restoration process via all-optical failure localization, we do not focus on novel FDP restoration capacity allocation schemes; instead, we would adopt state-of-the-art FDP SCA schemes for the proof of concept and facilitation of performance analysis. To be specific, the ILP in [14] and *Successive Survivable Routing (SSR)* algorithm in [10] are implemented in Sect. 5.6 for comparison with other counterparts as well as for understanding the performance behaviors under the monitoring resource hidden property, respectively.

5.4 The Monitoring Resource Hidden Property

With the SCA problem formulated in the previous section, we are particularly interested in the monitoring resource hidden property, which defines and quantifies in what circumstances the monitoring WLs are all reused by the P-LPs. In other words, there is no additional resource consumed due to the deployment of the m-trails and thus the spare capacity consumption is the same as that of the conventional FDP which is optimal among all the protection schemes. As a first study into the proposed framework [18], we analyze the problem under single-link SRLGs in circulant graphs, aiming to gain deeper understanding on the performance behavior of the proposed framework.

5.4.1 Lower Bound on the Spare Capacity

Lemma 5.1. *Let K be a set of links that form a cut in G. The spare capacity along the links in K is at least*

$$\frac{1}{|K|-1} \sum_{e \in K} q_e \le \sum_{e \in K} p_e. \tag{5.3}$$

Proof. In case link e fails, protection routes must be able to circumvent e via the other links in K, thus the spare capacity along those links is at least q_e, formally

$$q_e \le \sum_{f \in K, e \ne f} p_f. \tag{5.4}$$

There are $|K|$ such inequalities. By summing up these inequalities we get

$$\sum_{e \in K} q_e \le \sum_{e \in K} \sum_{f \in K, e \ne f} p_f = (|K|-1) \cdot \sum_{e \in K} p_e. \tag{5.5}$$

Finally, dividing both sides by $(|K|-1)$ we get Eq. (5.3).

Corollary 5.1. *Let $K^1, \ldots, K^k$ be pairwise disjoint cuts. Then*

$$\mathcal{M} \ge \sum_{1 \le l \le k} \sum_{e \in K^l} \frac{q_e}{|K^l|-1}. \tag{5.6}$$

Corollary 5.2. *Let $N_G(v)$ denote the set of links adjacent to node v and $\Delta_v = |N_G(v)|$*

$$2\mathcal{M} \ge \sum_{v \in v} \sum_{e \in N_G(v)} \frac{q_e}{\Delta_v - 1}. \tag{5.7}$$

As a rule of thumb, the minimal ratio of spare to working capacity can be estimated by $1/(\overline{\Delta}-1)$ [8], where $\overline{\Delta} = 2|E|/|V|$ is the average nodal degree. It was proved for the case when the working capacity is the same on every link [8]. This can be also deduced from Corollary 5.2 by applying the inequality of arithmetic and harmonic means.

5.4.2 Dominance of Monitoring Resources

With the proposed NL-UFL construction in the $C_n(1,2)$ topologies, we are interested in whether and how much monitoring resources can be hidden by restoration resources. Let the average working capacity per link, the average restoration capacity per link, and average monitoring capacity per link be denoted by

$$\bar{q} = \frac{1}{|E|}\sum_{e \in E} q_e, \quad \bar{p} = \frac{1}{|E|}\sum_{e \in E} p_e, \quad \bar{m} = \frac{1}{|E|}\sum_{e \in E} m_e,$$

respectively. Formally, the *dominance of monitoring resources* occurs when $\bar{m} \geq \bar{p}$, which serves as a sufficient condition that additional monitoring resources are required by FDP on top of the restoration capacity.

Let θ measure the traffic demand as a percentage of $s-d$ pairs that are loaded with a W-LP. For example, $\theta = 100\%$ means each $s-d$ pair (i.e., $|V|(|V|-1)/2$) is connected by a W-LP. It is clear that with smaller θ, the dominance of monitoring resources is more likely to happen. There naturally comes up an interesting question: with a specific topology, for which values of θ will make the monitoring resources dominant? Is there a lower bound on θ, say 1%, below which $\bar{m} < \bar{p}$ is unconditionally true? In the following theorem we show that there is no such lower bound on θ in a circulant topology $G = C_n(1,2)$ under single-link SRLGs.

Theorem 5.1. *For any $0 < \theta < 1$, there exists an FDP-SCA problem (with topology, SRLG and traffic) where the monitoring resources will never dominate the spare capacity (i.e., $\bar{m} < \bar{p}$).*

Proof. We pick the circulant graphs $G = C_n(1,2)$ as a topology with unit cost along each link, and all single-link failures for SRLGs. The nodes of $G = C_n(1,2)$ are $0, 1, \ldots, n-1$, and the edges are $(v, v+1)$ and $(v, v+2)$, for $v = 0, \ldots, n-1$, where the addition is understood modulo n. Let us call the edges $(v, v+1)$ on-cycle edges and the rest chordal edges (see also Fig. 3.10). As for the traffic, let G be launched with a set of shortest-path routed W-LPs denoted as $\mathscr{W} = W_1, \ldots, W_k$, such that $k \leq \theta \cdot \frac{n(n-1)}{2}$. Let $h = \max\{5, \lceil\frac{1}{\theta}\rceil\}$ and $n = 4h \geq 20$. Nodes s and d are connected with a W-LP along the shortest path route if $s-d \equiv 0 \mod h$. In this case the number of undirected connections is $k = \frac{3}{2}n$, because each source node s is connected with $s+h$, $s+2h$, and $s+3h$, where addition is understood modulo n. Therefore, the total number of undirected connections is

$$\theta \cdot \frac{n(n-1)}{2} = \theta \cdot 4 \cdot \max\{5, \lceil\frac{1}{\theta}\rceil\}\frac{n-1}{2} \geq 2(n-1) \geq \frac{3}{2}n = k$$

Next, we define a set of disjoint cuts $K^1, \ldots, K^{n/4}$, where K^i contains six links $(v-1, v+1)$, $(v, v+1)$, $(v, v+2)$ for both $v = 2i$ and $v = 2i + n/2$. Each cut K^i, $i = 1, \ldots n/4$ separates the graph into two equal-size fragments with $\frac{n}{2}$ nodes, thus the number of W-LPs passing through the cut is at least n, i.e., $\sum_{e \in K^l} q_e \geq n$, because for each source node s the number of possible destination nodes in the other side of the cut is 2, thus there are at least n for which $s - d \equiv 0 \mod h$ holds.

According to Corollary 5.1 we have $\mathcal{M} \geq \frac{1}{5} \cdot n \cdot \frac{1}{4} n$. Note that, the number of links $|E| = 2n$, thus $\overline{p} = \frac{\mathcal{M}}{2n}$. According to Theorem 3.7, the total monitoring capacity of the bm-trails is $n \lceil \log_2(2n + 1) \rceil + 1$. Thus we have $\overline{m} = \frac{1}{2} \lceil \log_2(2n + 1) \rceil + \frac{1}{2n}$. Therefore, $\overline{m} < \overline{p}$ holds when $\frac{1}{2} \lceil \log_2(2n + 1) \rceil + \frac{1}{2n} < \frac{1}{40} \cdot n$. The left-hand side is a logarithmic function, thus it is always true for large enough n.

It is important to note that although necessary, the dominance of monitoring resources is not sufficient for nonzero monitoring overhead.

5.5 General Topologies with Multi-link SRLGs

An m-trail allocation scheme for NL-UFL under multi-link SRLGs, called *Network-Wide Local Link Code Construction (NL-LCC)* reported in [19], is presented in this section.

The basic idea of NL-LCC is to successively and incrementally construct the alarm code table at each node. A similar idea was explored in LCC [1] in Sect. 3.3.6; however, different from any previously reported scheme, NL-LCC meets all the desired features of the proposed framework which incorporates in-band information, deals with multi-link SRLGs, and can achieve NL-UFL.

Let $\underline{\underline{A}}^v$ at node v denote the ACT of v, which is a binary matrix on all the m-trails traversing through v, with each entry $a^v_{z,[j]} = 1$ if T_j passes SRLG z, $a^v_{z,[j]} = 0$ if T_j is disjoint from z, and "x" (i.e., a *don't care bit*) if it does not affect the NL-UFL property by having T_j pass through z or not. The pseudo code of NL-LCC is given in Algorithm 5 and is explained step by step as follows.

In Step (3), each node has an initial ACT generated based on the terminated W-LPs, where $a^v_{z,[j]} = 1$ if W_j terminated at node v passes through z, and 0 otherwise. Then, the heuristic enters the loop in Steps (4)–(5) which checks each pair of SRLGs to ensure their alarm codes are different and distinguishable at any node v. The objective of the heuristic is that all the SRLG alarm codes are unique by flipping as few don't care bits to 0 or 1 as possible.

If two SRLG z_1 and z_2 are found with a common alarm code (i.e., $a^v_{z_1} = a^v_{z_2}$ in Step (6)), the heuristic first checks if the last bit of each alarm code is x (Step (7)). If not, in Step (8) one more bit is appended and tapped with a don't care bit x, where the length of alarm codes at v is increased by one (i.e., $b_v = b_v + 1$).

Note that ensuring the last bit position of SRLG alarm code as x is necessary for the guaranteed success in distinguishing the alarm code of SRLG z_1 and z_2 through Steps (10)–(23).

In Step (11) we iterate through each link $e \in (z_1 \cup z_2) \setminus (z_1 \cap z_2)$ from the symmetric difference of z_1 and z_2 (e.g., $e \in z_1$, $e \notin z_2$) and in Step (12) we iterate through every bit position j. In Step (13) if the jth bit of the alarm code of e is x

5.6 Performance Evaluation

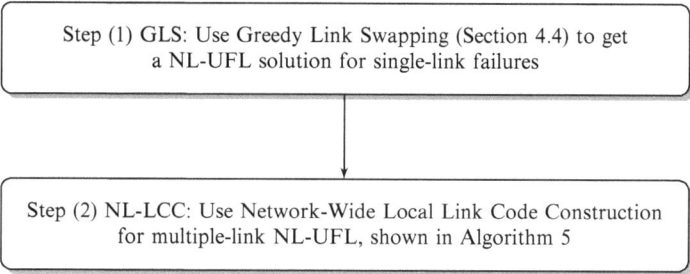

Fig. 5.3 The flowchart for the proposed multi-link SRLG NL-UFL algorithm using both in-band and out-of-band information in Step (2)

or 1, and $\forall f \in z_2$ the jth position is x or 0, we treat link e and position j as a good candidate for distinguishing the failure of SRLG z_1 and z_2. Next, in Step (14) we evaluate the cost of this candidate, which is the number of don't care bits among $a^v_{e,[j]}$, and $a^v_{f,[j]}$, $\forall f \in z_2$. Our goal is to select the possible link e and position j where the fewest number of bits should be set to 0 or 1, and thus most don't care bits remain for the next iterations, which is stored in the working variables e_m and j_m in Step (16). Note that, such a bit position always exists because the last bit is x due to Step (8). Next, the best candidate is selected in Step (21), and bit j_m of link e_m is set to 1, while $\forall f \in z_2$ it is set to 0 at position j_m. In such a way we can ensure $a^v_{z_1,[j_m]} \neq a^v_{z_2,[j_m]}$, which will remain unchanged regardless of the future iterations.

Finally, we need to make sure that all the information is local at node v; thus, we search for the shortest path in G between node v and the closest terminal node of e_m through links f with $a^v_{f,[j_m]} = \{1, x\}$ in Step (22), and set those bits to 1 in Step (23).

The proposed heuristic for multi-link SRLG NL-UFL is composed of two parts implemented one after the other, shown in Fig. 5.3. A set of bm-trails are first routed by using Greedy Link Swapping (GLS) (Sect. 4.4) to distinguish all the single-link SRLGs; then another set of bm-trails are allocated in order to distinguish dual-link SRLGs as well with NL-LCC in Algorithm 5. An illustration of the additional m-trails allocated in Step (2) of Fig. 5.3 is presented in Fig. 5.4.

5.6 Performance Evaluation

5.6.1 Comparison of Signaling-Free Protection Methods

First, a case study is presented to examine the proposed framework in terms of (1) capacity efficiency, (2) restoration time, and (3) computation time. We launched the m-trails to achieve NL-UFL and implemented an FDP scheme (referred to as FDP) as a realization of the proposed framework, which is compared to a couple of signaling-free restoration schemes, namely Collaborative Failure Protection [22]

Algorithm 5: Network-Wide Local Link Code Construction (NL-LCC) algorithm

Input: $G = (v, E), \mathscr{Z}, W_1, \ldots, W_k$
Result: Set of $T_1, \ldots, T_b$ bm-trails
1 **begin**
2 | **for** $v \in V$ **do**
3 | | Construct an initial ACT at node v.
4 | | **for** $z_1 \in \mathscr{Z}$ **do**
5 | | | **for** $z_2 \in \mathscr{Z}$ **do**
6 | | | | **if** $z_1 \neq z_2 \wedge A_{z_1}^v = A_{z_2}^v$ **then**
7 | | | | | **if** $\exists e, a_{e,b_v}^v = \{0,1\}$ **then**
8 | | | | | | Add a new bit x to each alarm code
9 | | | | | **end**
10 | | | | $e_m, j_m, d_m = \infty$
11 | | | | **for** $e \in (z_1 \cup z_2) \setminus (z_1 \cap z_2)$ **do**
12 | | | | | **for** $j = 1, \ldots, b_v$ **do**
13 | | | | | | **if** $a_{e,j}^v = \{1,x\} \wedge a_{f,j}^v = \{0,x\}, \forall f \in z_2$ **then**
14 | | | | | | | $d :=$ number of x bits in $a_{e,j}^v, a_{f,j}^v, \forall f \in z_2$.
15 | | | | | | | **if** $d < d_m$ **then**
16 | | | | | | | | $d_m = d, e_m = e, j_m = j$
17 | | | | | | | **end**
18 | | | | | | **end**
19 | | | | | **end**
20 | | | | **end**
21 | | | | Set $a_{e_m,j_m}^v = 1$ and $a_{f,j_m}^v = 0, \forall f \in z_2$.
22 | | | | **Find the shortest path** between node v and link e_m along links with don't care or 1 bits at position j_m.
23 | | | | **Set** the don't care bits along the path to 1.
24 | | | **end**
25 | | **end**
26 | | **end**
27 | **end**
28 **end**

(referred to as CFP) and p-Cycle. Both p-cycle and CFP organize spare capacity into pre-cross-connected cycles to achieve signaling-free protection. CFP can achieve better capacity efficiency by reusing the released working capacity of the disrupted lightpaths (i.e., stubs). It is achieved in a cooperative manner among the failure-aware nodes. We computed the optimal solutions by implementing and solving the ILP for each of the three schemes, specifically the one in [22, 23] and [14, Chap. 9.5.2] for p-Cycle, CFP, and FDP, respectively, in which the same link cost and traffic values were adopted to ensure the experiment environments completely in line with the previous art. We assume the given traffic matrix corresponds to 1 Erlang; each link has 16 WLs, and the W-LPs are shortest-path routed in each case. The results were obtained by solving the ILPs using CPLEX v.11 with zero optimality gap.

5.6 Performance Evaluation

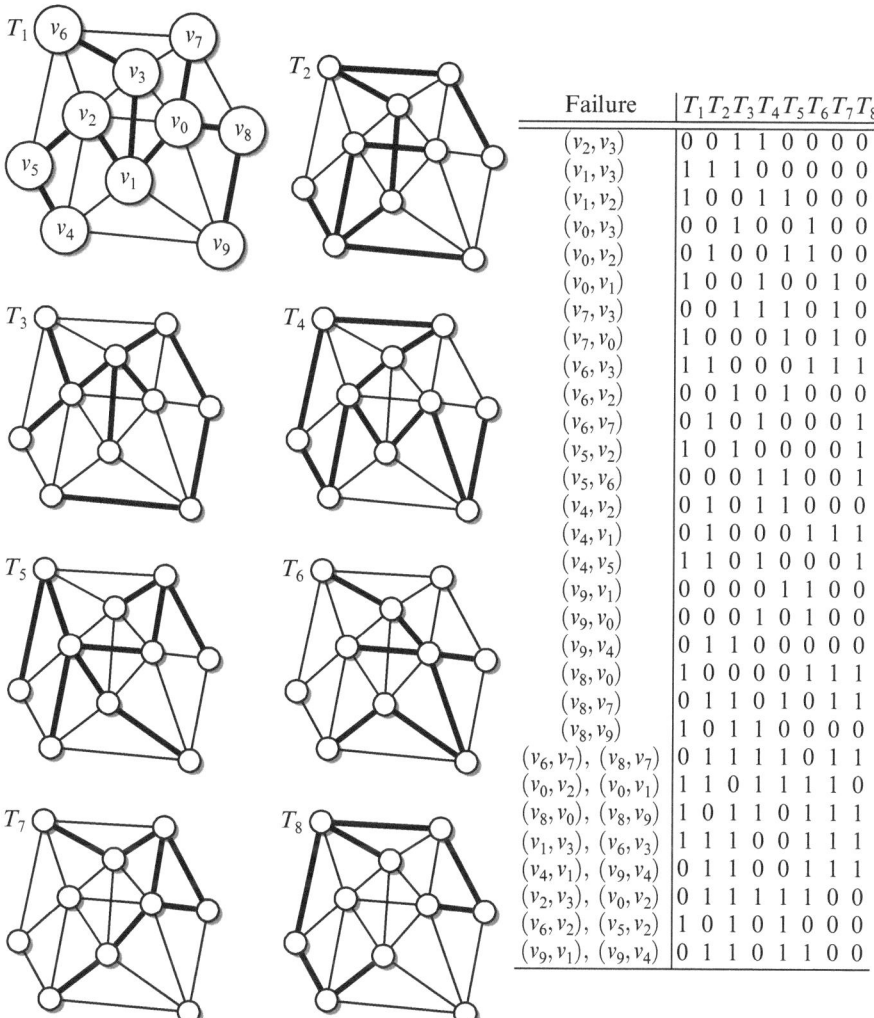

Fig. 5.4 An NL-LCC solution based on bm-trails. The solution has $b = 8$ and $\|\mathscr{T}\| = 67$

5.6.1.1 Capacity Efficiency

Figure 5.5 compares the capacity efficiency of the three schemes. It clearly shows that the performance of the proposed FDP is dominated by the monitoring resource consumption by the m-trails when the traffic load is slight, but becomes the most efficient one when the traffic load is increased because all the monitoring resource consumption can be hidden by the spare capacity taken by the P-LPs of the FDP scheme. Note that FDP yields the optimal capacity efficiency among all the

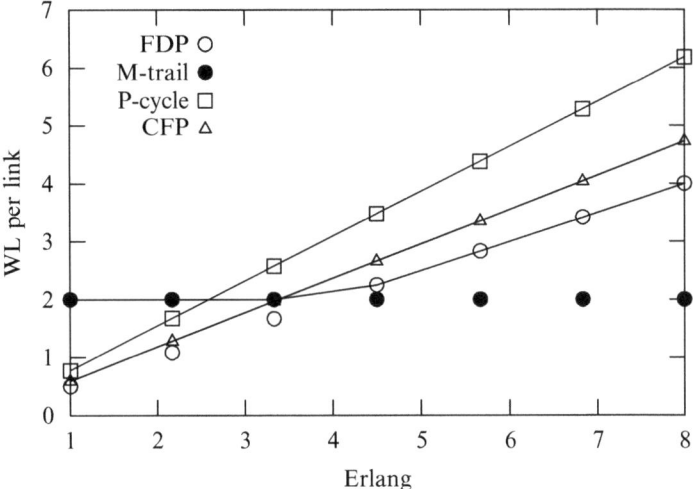

Fig. 5.5 WLs per link on Smallnet

protection schemes, and the capacity consumption by FDP is completely dominated by the spare capacity when the traffic load is heavy enough to hide all the monitoring resource consumption.

5.6.1.2 Restoration Time

We assume the average physical length of a link is 200 km, which requires 1 ms for light to traverse; thus, the failure localization and notification delays under the FDP and NL-UFL based protection framework are $t_l + t_n < 20$ ms since the longest m-trail is no longer than 20 hops. By assuming $t_c + t_p = 10$ ms and $t_d = 20$ ms, this leads to an overall average restoration time of 50 ms. For CFP and p-Cycle, the maximal length of each cycle is 10 hops, which yields $t_l < 10$ ms for failure localization and notification; however, we also have $t_d = 20$ ms required at the two nodes for switch fabric configuration and $t_c + t_p = 10$, which results in 40 ms in total. For CFP, the longest p-Cycle has seven links, which leads to slightly shorter restoration time than 40 ms. Under the specification of the case study, the three schemes lead to very similar restoration time.

5.6.1.3 Computation Time

The computation time was 39.00 s, 17.22 s, and 3.94 s for CFP, p-Cycle, and FDP, respectively. Thus, we can conclude that FDP has the best computation efficiency.

5.6.1.4 Under Multi-link SRLGs

We have adopted the problem instance in [17] for comparison, which is, to the best of our knowledge, the only previously reported study that provides an ILP for static multi-link SRLG p-cycle design. Note that all the other studies for multi-link SRLGs using p-cycles have focused on reconfiguration, rerouting, and dynamic reconfiguration of spare capacity, which do not fit into the targeted scenario. The NSF network (14 nodes, 22 links) topology is used for all single-link and double-link SRLGs. P-cycle requires 965 bandwidth units while the NL-UFL and FDP framework requires 302. This clearly shows that the proposed NL-UFL and FDP approach can gain even more advantage when multi-link SRLGs are considered. The average length of m-trails is 20 links, while the maximum is 26 links, which still results sub 50 ms restoration time for every single and double failure with NL-UFL and FDP approach.

5.6.2 Monitoring Resources Hidden

We have seen that the proposed approach is outperformed by its counterparts when the working traffic load is small, but will become much more efficient when the number of W-LPs increases, and the monitoring resources can be completely hidden by the restoration capacity for FDP. Therefore, we claim that the proposed approach can achieve the same capacity efficiency as conventional FDP provided the network is loaded with sufficiently large working capacity. This subsection provides extensive simulation results in a wide range of network topologies and SRLG densities, so as to gain deeper understanding on the monitoring resource hidden property under the proposed restoration framework.

The randomly generated planar graphs are classified according to parameter g, which is the length of the longest inner face contained in the graph. Clearly, graphs with smaller values of g are considered more densely meshed; and in the generated graphs, the average nodal degree of each graph ranges from 5.4 (for $g = 3$) to 2.76 (for $g = 7$). See also Fig. 1.1j–l for example network topologies with different g. Besides, the COST266 European reference network (37 nodes, 57 links) is also used [11]. We adopted SRLGs with single-link and double adjacent links, and the SRLG density considered in the problem is parameterized by a *double failure density parameter*, denoted by f, which indicates the fraction f of all double adjacent link SRLGs under consideration.

We evaluate the resource consumption in the proposed framework by increasing the working traffic, which is defined as the percentage of $s - d$ pairs that are interconnected by a W-LP (i.e., θ). Let every W-LP and m-trail take a single wavelength of bandwidth (i.e., a single WL). The ILP in [14] was no longer used here due to the huge computational complexity; instead, we adopted SSR [10] that sequentially allocates P-LPs for each W-LP. SSR iteratively launches Dijkstra's shortest path algorithm to find disjoint backup paths with each SRLG involved in a W-LP one at a time using the latest spare capacity information.

Figure 5.6a, b show the average monitoring overhead, i.e.,

$$\frac{1}{|E|} \cdot \sum_{e \in E} \max\{0, m_e - p_e\},$$

as θ is increased from 0 to 100 %. It meets our expectation that as the percentage of loaded $s - d$ pairs (i.e., θ) is decreased, and/or as the percentage of double-link adjacent SRLGs (i.e., f) increases, the monitoring overhead increases accordingly. It is interesting to observe that the topology density does not affect the monitoring overhead as shown in Fig. 5.6b, because taking a sparser topology increases the total monitoring capacity for m-trails, meanwhile increasing the required restoration capacity for FDP, too. As a rule of thumb, we claim that the monitoring overhead is negligible if at least 50 % of the node-pairs in a network are loaded with a W-LP.

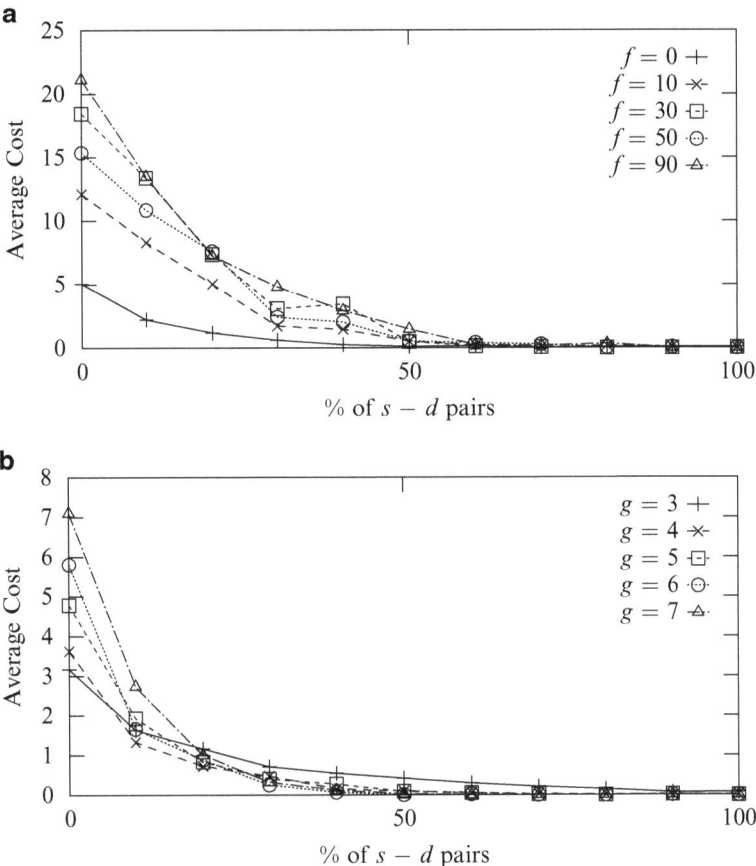

Fig. 5.6 Average monitoring overhead. (**a**) COST266 network, (**b**) random networks with $f = 0$

5.6 Performance Evaluation

Further, with more than 20 % of loaded $s - d$ pairs, the monitoring overhead is 1 WL per link under single-link SRLGs, regardless of the network density.

In Fig. 5.7a, b, the average numbers of WLs per link for W-LPs, P-LPs, and m-trails are evaluated in the COST266 European reference network shown on Fig. 1.1h (37 nodes, 57 links) with $f = 0$ and $f = 90\%$ in (a) and (b), respectively. The intersections of the curves for average restoration capacity per link (i.e., $\overline{p}$) and average monitoring capacity per link (i.e., $\overline{m}$) are at $\theta = 20\%$ and $\theta = 40\%$ with an SRLG density $f = 0\%$ and $f = 90\%$, respectively. The curve of $\overline{m}$ shows the contribution adopting in-band information in the reduction of monitoring resource consumption, which is not obvious in the single-link SRLG scenario in Fig. 5.7a. Nonetheless, the effect of taking in-band information becomes nontrivial when multi-link SRLGs are considered, where $\overline{m}$ decreases from 21.1 to 13.5 as

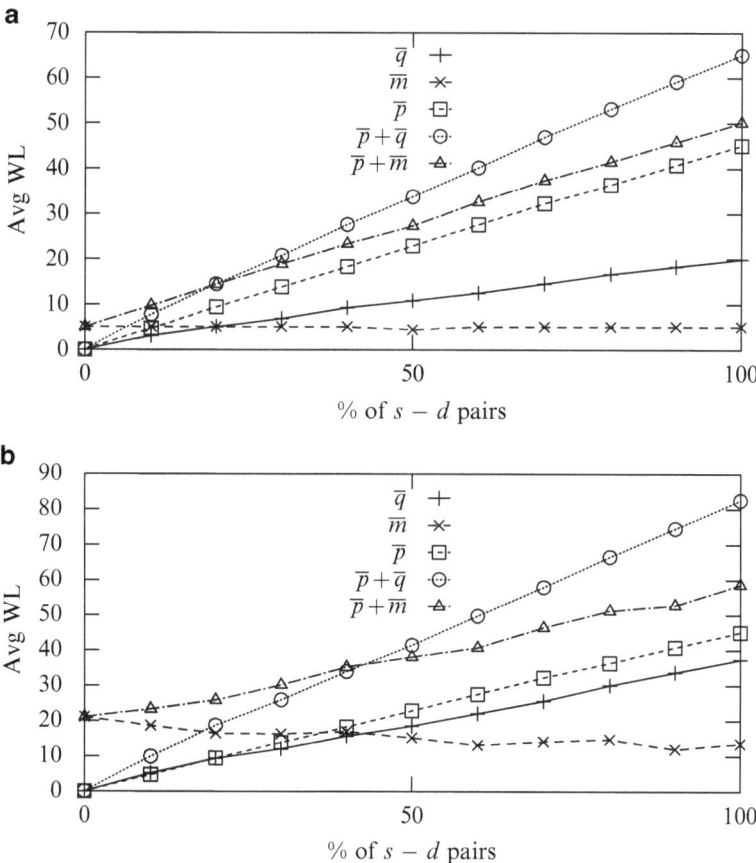

Fig. 5.7 Average used capacity on the links in the COST266 reference network, where working, spare, and m-trail capacity is denoted by $\overline{q}$, $\overline{p}$, and $\overline{m}$, respectively. (**a**) Single-link SRLGs ($f = 0\%$). (**b**) Adjacent double-link SRLGs ($f = 90\%$)

θ increases from 0 % to 100 %, as shown in Fig. 5.7b. Further, the reduction of $\overline{m}$ occurs mostly in the light traffic region where the monitoring resources are more likely to be dominant.

Figure 5.8a, b show the relation between the maximum total capacity required along each link (i.e., $\overline{q} + \max\{\overline{m}, \overline{p}\}$) and θ by using COST266 European reference network (37 nodes, 57 links), under various SRLG densities and topology densities, respectively. In Fig. 5.8a when $\theta = 50\%$, the required maximum WLs is roughly 80. This provides us a design guideline that with 80 WLs of total capacity consumed along each link in COST266, the proposed framework can most likely achieve optimal performance, due to the fact that $\theta = 50\%$ is found to be the turning point for reaching zero monitoring overhead under multi-link SRLGs, as shown in Fig. 5.6a. Further, as shown in Fig. 5.8b, 30–45 WLs of overall capacity along each

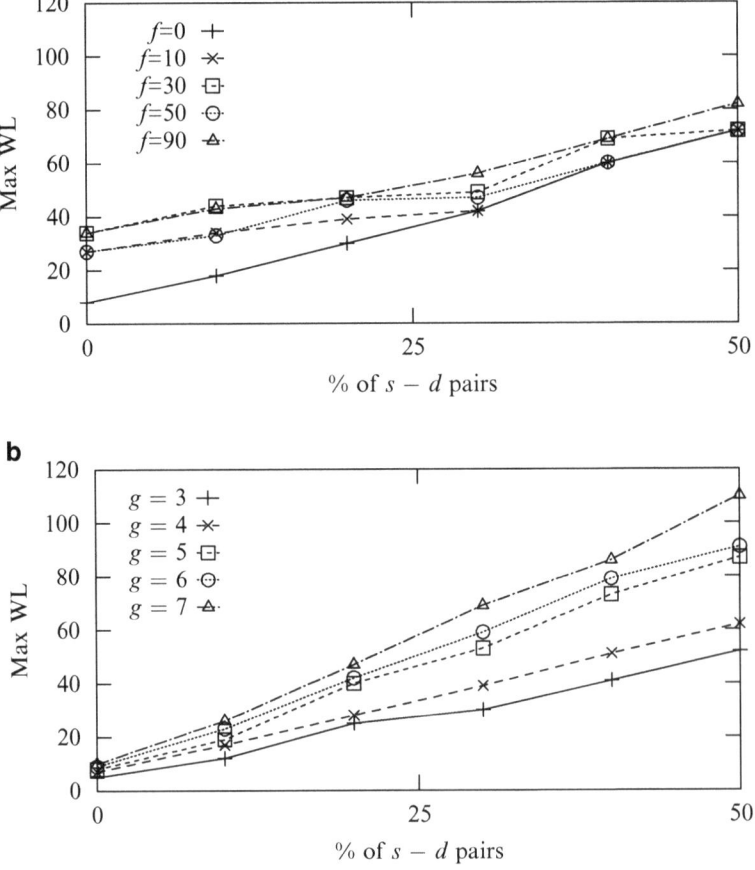

Fig. 5.8 Maximum overall capacity in WLs. (**a**) COST266 network. (**b**) Random networks with $f = 0$

link is the threshold for zero monitoring overhead for single-link failures, because $\theta = 20\,\%$ is the turning point of zero monitoring overhead under single-link SRLGs, as shown in Fig. 5.6b.

5.7 Summary

This chapter introduced an approach that can enable a generic shared protection scheme, such as FDP, to achieve the following three desired features of optical layer protection in all-optical mesh networks: (1) a signaling-free and completely all-optical restoration process; (2) 100 % restorability for any multi-link SRLG failure event; and (3) optimal capacity efficiency as FDP under given conditions. By incorporating NL-UFL, a new problem formulation for SCA is given and is composed of two parts: monitoring resource allocation for m-trails, as well as restoration resource allocation for implementing FDP. We analyzed the formulated problem regarding its monitoring resource hidden property, where a polynomial time deterministic construction on circulant topologies was developed and proved that the dominance of monitoring resources does not occur even when traffic load (i.e., θ) is close to zero.

The chapter also presented a heuristic approach, called NL-LCC, for solving the formulated problem in general topologies. A comparison between NL-LCC and two of its counterparts, namely p-Cycle and cooperative fast protection, was made in terms of capacity efficiency. Some observations were gained as follows: (1) the NL-LCC scheme outperforms the other two schemes on capacity efficiency when working traffic is over some threshold, and such advantage is getting more significant when there is more working traffic; (2) a turning point exists in terms of the percentage of loaded $s-d$ pairs (i.e., θ) which results in zero monitoring overhead. The turning point is at $\theta = 50\,\%$ and $\theta = 20\,\%$ for multi-link SRLGs and single-link SRLGs, respectively; (3) a rule of thumb was identified to estimate whether NL-LCC achieves zero monitoring overhead using the amount of maximum occupied capacity along each link.

References

1. Babarczi P, Tapolcai J, Ho PH (2011) SRLG failure localization with monitoring trails in all-optical mesh networks. In: Proceedings of the international workshop on design of reliable communication networks (DRCN). IEEE, Krakow, pp 188–195
2. Choi H, Subramaniam S, Choi HA (2004) Loopback recovery from double-link failures in optical mesh networks. IEEE/ACM Trans Netw 12(6):1119–1130
3. Doucette J, Grover WD (2001) Comparison of mesh protection and restoration schemes and the dependency on graph connectivity. In: Proceedings of the international workshop on design of reliable communication networks (DRCN). IEEE, Budapest, Hungary, pp 121–128

4. Frederick MT, Datta P, Somani AK (2006) Sub-graph routing: a generalized fault-tolerant strategy for link failures in wdm optical networks. Elsevier Comput Netw 50(2):181–199
5. Grover WD (2003) Mesh-based survivable networks: options and strategies for optical, MPLS, SONET and ATM networking. Prentice Hall PTR, Upper Saddle River
6. Grover W, Doucette J, Clouqueur M, Leung D, Stamatelakis D (2002) New options and insights for survivable transport networks. IEEE Commun Mag 40(1):34–41
7. Iraschko RR, Grover WD (2000) A highly efficient path-restoration protocol for management of optical network transport integrity. IEEE J Sel Areas Commun 18(5):779–794
8. Iraschko RR, MacGregor M, Grover WD (1998) Optimal capacity placement for path restoration in STM or ATM mesh survivable networks. IEEE/ACM Trans Netw 6(3):325–336
9. Liu Y, Tipper D (2001) Spare capacity allocation for non-linear link cost and failure-dependent path restoration. In: Proceedings of the international workshop on design of reliable communication networks (DRCN)
10. Liu Y, Tipper D, Siripongwutikorn P (2005) Approximating optimal spare capacity allocation by successive survivable routing. IEEE/ACM Trans Netw 13(1):198–211
11. Maesschalck S, Colle D, Lievens I, Pickavet M, Demeester P, Mauz C, Jaeger M, Inkret R, Mikac B, Derkacz J (2003) Pan-European optical transport networks: an availability-based comparison. Photonic Netw Commun 5(3):203–225
12. Martin R, Menth M, Canbolat K (2006) Capacity requirements for the one-to-one backup option in mpls fast reroute. In: Proceedings of the IEEE BroadNets, San Jose
13. Pan P, Swallow G, Atlas A (2001) Fast reroute extensions to rsvp-te for lsp tunnels. IETF RFC 4090
14. Pióro M, Medhi D (2004) Routing, flow, and capacity design in communication and computer networks. The Morgan Kaufmann series in networking, Morgan Kaufmann Publishers/Elsevier
15. Ramamurthy S, Mukherjee B (1999) Survivable WDM mesh networks, part ii - restoration. In: Proceedings of the IEEE ICC, pp 2023–2030
16. Ramasubramanian S, Harjani AS (2006) Comparison of failure dependent protection strategies in optical networks. Photonic Netw Commun 12(2):195–210
17. Sebbah S, Jaumard B (2009) P-cycle based dual failure recovery in WDM mesh networks. In: Proceedings of the IFIP working conference on optical network design & modelling (ONDM)
18. Tapolcai J, Ho PH, Rónyai L, Wu B (2012) Network-wide local unambiguous failure localization (NWL-UFL) via monitoring trails. IEEE/ACM Trans Netw 20(6):1762–1773
19. Tapolcai J, Ho P-H, Babarczi P, Rónyai L (2014) On signaling-free failure dependent restoration in all-optical mesh networks. IEEE/ACM Trans Netw (in press)
20. Wang D, Li G (2008) Efficient distributed bandwidth management for mpls fast reroute IEEE/ACM Trans Netw 16(2):486–495
21. Wang H, Modiano E, Médard M (2002) Partial path protection for wdm networks: end-to-end recovery using local failure information. In: Proceedings of the IEEE symposium on computers and communications (ISCC), pp 719–725
22. Wu B, Ho PH, Yeung KL, Tapolcai J, Mouftah HT (2010) Cfp: cooperative fast protection. IEEE/OSA J Lightwave Technol 28(7):1102–1113
23. Wu B, Yeung KL, Ho PH (2010) Ilp formulations for p-cycle design without candidate cycle enumeration. IEEE/ACM Trans Netw 18(1):284–295
24. Xiong Y, Mason L (1999) Restoration strategies and spare capacity requirements in self-healing atm networks. IEEE/ACM Trans Netw 7(1):98–110

Chapter 6
Global Neighborhood Failure Localization

Abstract The distributed m-trail framework (i.e., NL-UFL) introduced in the previous chapter enables every node to be able to instantly localize a failed SRLG such that the failure restoration can be performed automatically in the optical domain. This is obviously not necessary since a node may not need to respond to a failure event if the node is not traversed by any P-LP whose W-LP is subject to the failure. In response to such an observation, this chapter defines and investigates an interesting m-trail scenario, called Global Neighborhood Failure Localization (G-NFL). As a one step advance of NL-UFL, G-NFL defines the *neighborhood* of a node, which is a set of links whose failure states should be known to the node in restoration of the corresponding W-LPs, and the G-NFL problem routes a set of m-trails such that each node can localize any failure in its neighborhood. To gain insight into the G-NFL problem, the chapter provides bound analysis on the minimum bandwidth required for m-trails, along with a simple yet effective heuristics.

6.1 Introduction

NL-UFL assumes each node to be able to unambiguously identify all possible failures. Thus a node will monitor a remote link even if the node does not need to respond to the link failure, resulting in unnecessary monitoring resource consumption, high computation complexity, and very lengthy m-trails. As a remedy to save the monitoring resources, this chapter investigates an on-demand m-trail allocation paradigm that enables a general shared protection scheme to perform signaling-free failure restoration as in 1+1 and p-Cycle, called Global Neighborhood Failure Localization (G-NFL). The features of G-UFL are summarized as follows:

1. The *neighborhood* of a node is defined as a set of links whose failures must be unambiguously localized by the node, such that each node only localizes the link failures in its neighborhood.
2. The spare capacity by P-LPs can be reused to support the m-trails in order to achieve better capacity efficiency.
3. A node can monitor both traversing m trails and W LPs for failure status acquisition, which is referred to as *out-of-band* and *in-band* monitoring, respectively.

172 6 Global Neighborhood Failure Localization

4. Since each m-trail is at most one hop longer than the corresponding shortest path, a provisioned lightpath can be taken as an m-trail or W-LP or both according to the operation requirement.

The G-NFL problem is defined as to finding a set of m-trails with minimum consumed WLs (or called *coverlength*) such that all the nodes can localize the failures in their respective neighborhoods. Lower bounds of the G-NFL problem are derived for general graphs by using Combinatorial Group Testing (CGT) arguments. This is a newly defined CGT problem where the cost of testing a group is dependent on the size of the group, and the developed theorem and proof are considered the first theoretical work under the associated application scenario different from all the previously reported research.

6.2 The G-NFL Scenario

The *neighborhood* of a node is defined as a set of links whose failures must be unambiguously localized by the node. In particular, the neighborhood of a node should contain all the links along the W-LPs whose corresponding P-LPs traverse through the node. On the other hand, all the nodes traversed by a P-LP should be able to localize the link failure for which the P-LP is used to restore. Therefore, the size of neighborhood of each node (and the resultant monitoring resource consumption) is expected to scale well with the network sizes.

6.2.1 Introduction of G-NFL

G-NFL is introduced as a novel scenario of m-trails aiming to resolve all the potential issues in the previously reported studies. Given the W-LPs and P-LPs, the neighborhood of each node is defined, and a node is said to meet the *NFL requirement* if it can localize all the link failure in its neighborhood. A feasible G-NFL solution consists of a set of m-trails such that each node can localize the failed link in its neighborhood based on the on–off status of a subset of the m-trails and/or W-LPs that pass through the node.

Figure 5.2 shows an example of G-NFL in a topology with four nodes and two W-LPs denoted as W_1 and W_2, each being provisioned with two physical lightpaths on the same route in opposite directions. W_1 is protected by two P-LPs, namely $P_1^{(v_3,v_4)}$ for link failure (v_3, v_4), and $P_1^{(v_4,v_1)}$ for (v_4, v_1) as in Fig. 5.2a, while W_2 is protected by a single P-LP denoted as P_2^* as in Fig. 5.2b. To ensure signaling-free restoration for W_1 and W_2, v_1 should be able to unambiguously identify the failure of (v_4, v_1) and (v_2, v_1), such that W_1 (or W_2) can be switched over to $P_1^{(v_4,v_1)}$ (or P_2^*) when (v_4, v_1) (or (v_2, v_1)) fails. Thus, the neighborhood of v_1 must contain the two links (v_4, v_1) and (v_2, v_1). On the other hand, since v_2 is traversed by all the three P-LPs (i.e., $P_1^{(v_3,v_4)}$, $P_1^{(v_4,v_1)}$, and P_2^*), it needs to react to any failure upon W_1 or W_2. Therefore, the neighborhood of v_2 should contain (v_3, v_4), (v_4, v_1), and (v_2, v_1). Similarly, we can define the neighborhood of v_3 as (v_3, v_4), and that of v_4 as (v_2, v_1), (v_3, v_4), and (v_4, v_1).

To achieve the NFL requirement according to the above nodal neighborhoods, three m-trails T_1, T_2, and T_3 are needed as shown in Fig. 5.2c, by which the *alarm code table (ACT)* for each node is formed as shown in Fig. 5.2d–g. Each row of an ACT over the separator corresponds to a failure state within its neighborhood, and the rows below the separator are for the alarm codes seen at the node due to a link failure outside the neighborhood. All codes over the separator in an ACT should be unique. For example, v_1 keeps the ACT as in Fig. 5.2f by observing the on–off status of T_1, W_1 and W_2, so as to uniquely identify the failure on (v_4, v_1) or (v_2, v_1) in its neighborhood. If v_1 finds that T_1 and W_1 become unexpectedly off while W_2 is still on, an alarm code $[1, 1, 0]$ is obtained; so the node will consider link (v_4, v_1) as failed by matching the first row of its ACT and be ready to switch W_1 over to $P_1^{(v_4,v_1)}$. Meanwhile and in parallel, v_2 and v_4 will be able to identify the failure of (v_4, v_1) by matching the second row in their ACTs as in Fig. 5.2g and d, respectively, and instantly configure their OXCs to support $P_1^{(v_4,v_1)}$. Thus W_1 can be restored in an all-optical and deterministic fashion upon the failure of (v_4, v_1).

6.2.2 Resource Consumption by G-NFL

Resources for G-NFL are identified as *transponders* (or referred to as transmitters in the following context), *lambda monitors*, and *coverlength*.

6.2.2.1 Transmitters

Transmitters are expensive optical devices. However, we claim that the number of transmitters should not be an issue due to the following two reasons. Firstly, network providers usually prepare some amount of spare transmitters available at each OXC for heavy traffic loads. Secondly, in case more than the spare transmitters are required at a node, an optical splitter can be used to support multiple m-trails originated from the node.

6.2.2.2 Lambda Monitors

Lambda monitors have been built-in devices in commercial DWDM equipment such as OXCs and reconfigurable optical add-drop multiplexers (ROADMs) [3, 6] for the purpose of automatic power leveling. They are essential in adjusting the signal power of individual optical channels for all-optical amplification, mostly done by attaching a monitoring photo-diode at each channel port. Thus, very little cost is incurred due to the required lambda monitoring capability at each OXC.

6.2.2.3 Coverlength

Coverlength is the total number of WLs taken by the m-trails, which has been taken as the metric to evaluate the m-trail solutions [1, 4, 8, 10], and will still be the performance measure of this study.

6.2.3 Problem Definition

The input of the single-link G-NFL problem is an undirected graph $G = (V, E)$ with node set V and link set E, where the number of nodes is denoted by $n = |V|$ and the number of links by $m = |E|$. Given a set of W-LPs, denoted by $\mathscr{W}$, each of which can be in-band monitored by the nodes traversed by it for failure status acquisition. If a working path W_i that traverses e is interrupted due to the failure of e, the corresponding protection path P_i^e should be activated at the switching node for restoration. Let the *neighborhood* of node v be denoted by E_v, which is a set of links whose failure states can be unambiguously identified by v. Conversely, let *visibility region* of e be denoted by v_e, as a set of nodes each being able to unambiguously identify the failure of e.

The single-link G-NFL problem is to establish a set of m-trails to meet the two requirements (R1) and (R2) as follows.

(R1): each m-trail is a loopless path of G, at most one hop longer than the shortest path between its end nodes.

Here an m-trail is no longer than the minimum hop distance between the endpoints plus one, so that a provisioned lightpath can switch its role between a W-LP and m-trail according to the traffic demand and monitoring requirement. Such flexibility is desired in an intelligent failure localization framework and cannot be achieved with lengthy m-trails as in the previous studies [4, 8, 9]. Furthermore, short m-trails bear much better physical-layer impairment properties [5] than long m-trails.

The set of m-trails is denoted by $\mathscr{T} = \{T_1, \ldots, T_b\}$ where b is the number of m-trails. The objective is to minimize:

$$\|\mathscr{T}\| = \sum_{i=1}^{b} |T_i|, \qquad (6.1)$$

where $|T_i|$ is the number of links in m-trail T_i. We expect that each node $v \in V$ can achieve NFL according to the on–off status of m-trails and W-LPs in $\mathscr{T}^v$—the subset of $\mathscr{T} \cup \mathscr{W}$ containing the m-trails and W-LPs passing through v. Let $\underline{\underline{A}}^v$ denote the ACT at node v, where the ith bit of alarm code for link e at v will be denoted by $a_{e,i}^v$ for $1 \leq i \leq |\mathscr{T}^v|$, where $|\mathscr{T}^v|$ is the number of m-trails in $\mathscr{T}^v$. We have $a_{e,i}^v = 1$ if the ith m-trail passing through node v has link e and 0 otherwise.

6.2 The G-NFL Scenario

Table 6.1 M-trail reconfiguration upon dynamic data plane changes

	SOD-O	SOD-IO	LOD-O	LOD-IO
W-LP deployment	X	X		
W-LP release		X		X

To achieve NFL at node v, (R2) should follow:

(R2): every link e in neighborhood E_v has a unique nonzero alarm code seen at v denoted by A_e^v, and meanwhile different from all the possible link codes that v can see outside the neighborhood.

The following theorem proves the feasibility of the G-NFL problem in any connected graph.

Theorem 6.1. *Given a connected graph $G = (E, v)$ with neighborhoods E_v for every $v \in V$, an m-trail solution for G-NFL can always be found.*

Proof. One can use the argument of Theorem 1 in [8].

6.2.4 Neighborhood

An important feature of the G-NFL scenario is that each node only monitors the links in the neighborhood. The following two classes of neighborhood definitions are studied, while the dependencies between the monitoring and data plane are summarized in Table 6.1.

6.2.4.1 Strictly On-Demand

Strictly on-demand (SOD) enables a node v to failure-localize link e only if node v is involved in the restoration process of the link failure e *according to the current traffic distribution*; i.e., v is either the switching, intermediate, or merging node of a P-LP which protects link e along an active W-LP. It is expected to achieve the most efficient allocation of monitoring resources due to the SOD nature, but at the expense that the dynamic W-LP and P-LP route information has to be considered. Such strong dependency between monitoring and data planes imposes the need for reconfiguration of m-trails upon traffic distribution variations.

We consider two versions of SOD according to whether the W-LPs are taken for in-band monitoring or not, namely *SOD with out-of-band monitoring (SOD-O)*, and *SOD with in-band and out-of-band monitoring (SOD-IO)*. With the former, each node relies only on out-of-band monitoring for network failure status acquisition; while with the latter, a node can perform both out-of-band and in-band monitoring.

Obviously, both SOD-O and SOD-IO are subject to reconfiguration of m-trails upon any newly allocated W-LP. In the case of connection release, SOD-IO needs to check whether the W-LP is currently being used for in-band monitoring. If yes, the W-LP should be kept and automatically turned into an m-trail instead of being torn down immediately.

6.2.4.2 Loosely On-Demand

Loosely on-demand (LOD) aims to significantly reduce or completely avoid reconfiguration of m-trails by maintaining a clean separation between the monitoring and data planes. It defines the neighborhood of each node by *considering all the future possible traffic* (or when the network is *fully loaded* in the following context). Here, we define that the network is *fully loaded* when every node pair has at least one shortest-path-routed W-LP that is protected by one or a set of P-LPs under FDP; and if two W-LPs are allocated across a common node pair, they will be routed via a common route. Two versions of LOD are implemented, namely *LOD-O* and *LOD-IO*. The former considers only out-of-band monitoring, such that the m-trails can completely ignore the arrival and departure of the W-LPs. LOD-IO is different from LOD-O by taking in-band monitoring, which results in some extent of dependency between the monitoring and data planes; i.e., when a W-LP currently used for in-band monitoring is being released, the W-LP should be automatically turned into an m-trail for supporting the monitoring plane instead of being torn down immediately.

6.3 Bound Analysis

This section presents our bound analysis for the coverlength in the G-NFL problem. For the sake of simplicity and without loss of generality, let $\mathscr{W} = \emptyset$. We will first consider the lower bound on a generalized version of CGT and then apply them to the NFL requirement at each node, which will give us a lower bound on the coverlength for general graphs. The key idea is to define a special cost function for the m-trails at each node such that the lower bound to meet the NFL requirement at each node can be summed up to get a lower bound on the total coverlength.

6.3.1 Lower Bound for G-NFL

Theorem 6.2. *The total cover length for a G-NFL solution is at least*

$$\|\mathscr{T}\| \geq \sum_{v \in V} \left(1 - \frac{2}{m_v + 2}\right) \log_2(m_v) \quad (6.2)$$

6.4 G-NFL Heuristic

if $n - 1 \geq \frac{m_v}{2}$ for all $v \in v$, where m_v is the cardinality of the neighborhood of v, i.e., $m_v = |E_v|$.

Proof. We use an argument similar to the one in Sect. 4.3.3. By assumption, $n-1 \geq \frac{m_v}{2}$, thus we need to consider $x = |T_i| \leq \frac{m_v}{2} \leq n - 1$ only. Putting together the lower bounds on the cost in (4.25), (4.24) and applying Theorem 4.3 on each node we get a lower bound on Ω_v

$$\Omega_v \geq \min_{1 \leq x \leq \frac{m_v}{2}} \frac{2x}{1+x} \left(\log_2 x + \frac{m_v}{x} - 1 \right) \tag{6.3}$$

where inside the min there is a decreasing function for integer values of x by Lemma 4.1. Thus, it leads to

$$\Omega_v \geq \frac{2\frac{m_v}{2}}{\frac{m_v}{2}+1} \left(\log_2 \left(\frac{m_v}{2}\right) + \frac{m_v}{\frac{m_v}{2}} - 1 \right) =$$

$$= \frac{2m_v}{m_v + 2} (\log_2(m_v) - 1 + 2 - 1) =$$

$$= \left(2 - \frac{4}{m_v + 2} \right) \log_2(m_v). \tag{6.4}$$

Putting it together with (4.25), we get (6.2).

6.4 G-NFL Heuristic

The computational complexity of the optimal m-trail allocation problem with general multi-link failures is NP-hard [2], while being an open question for single-link failures (which is a special case of the multi-link failure scenario). Previous studies tackled the problem via heuristic methods that can approach the derived theoretical lower bounds [1, 7–9] in light of inefficiency of using integer linear programs to solve the problem. The chapter presents a simple yet effective heuristic to solve the G-NFL problem, where the ACT at each node is successively and incrementally constructed such that every link code is different from any other link code seen by the node.

The detailed description of the heuristic is given in Algorithm 6 and is explained step by step as follows.

In Step (2) an initial solution is taken by using single-hop m-trails for every link. The W-LPs $\mathcal{W}$ are given as the input of the algorithm ($\mathcal{W} = \emptyset$ in the -O schemes), and their visibility information is set in Step (3). In Step (4) each node $v \in V$ is considered one after the other to meet the NFL requirement such that each link code in the neighborhood E_v is unique. Specifically, E_v is loaded with W-LPs in Step (5), and the current ACT $\underline{\underline{A}}^v$ is constructed based on the m-trails traversing through node v in Step (6).

Then, the heuristic enters the loop in Steps (7)–(8) for each node v, by checking whether links e_1 and e_2, where $e_1 \in E_v$ and $e_2 \in E$, have the same alarm code seen at v or not. If yes in Step (9), we place an m-trail starting from v and traversing either e_1 or e_2, but not both. To make this information local at node v, we use Dijkstra's shortest path finding algorithm in Step (10) between v and the two adjacent nodes of the corresponding link, and select the one with the shortest distance in Step (11). Finally, we add the shortest possible path to $\mathcal{M}$ in Step (13) or Step (16), and refresh the ACT of v in Step (18).

The computational complexity of Algorithm 6 is described as follows. We have three iterations in Steps (4), (7), and (8) with $O(|V| \cdot |N| \cdot |E|)$ runs in total, where $|N|$ is the maximum size of the neighborhoods. In each iteration we compare the alarm codes, and compute a shortest path for any collided code pair with Dijkstra's shortest path finding algorithm, which can be done in $O(|E| + |V| \log |V|)$ steps. Altogether Algorithm 6 has $O(|V| \cdot |E| \cdot |N| \cdot (|E| + |V| \log |V|))$ steps in the worst

---

Algorithm 6: Global neighborhood failure localization

Input: $G = (V, E)$, W-LPs $\mathcal{W}$
Result: $\mathcal{T}$ set of m-trails
1 **begin**
2   Use a single-hop m-trail for each link as an initial guess;
3   Set W-LP visibility from $\mathcal{W}$;
4   **for** $v \in V$ **do**
5   Load the set of neighborhood links $E_v \subseteq E$;
6   Construct current ACT $\underline{\underline{A}}^v$ at node v;
7   **for** $e_1 = (u_1, w_1) \in E_v$ **do**
8   **for** $e_2 = (u_2, w_2) \in E$ **do**
9   **if** $e_1 \neq e_2 \wedge A^v_{e_1} = A^v_{e_2}$ **then**
10   Use Dijkstra's algorithm to get the shortest paths from v to $\{u_1, w_1, u_2, w_2\}$;
11   Set $\mathcal{P}_1$ and $\mathcal{P}_2$ to the shortest path to $\{u_1, w_1\}$ and $\{u_2, w_2\}$, respectively;
12   **if** $|\mathcal{P}_1| \leq |\mathcal{P}_2|$ **then**
13   Add m-trail $\forall e \in \mathcal{P}_1 \cup e_1$ to $\mathcal{T}$;
14   **end**
15   **else**
16   Add m-trail $\forall e \in \mathcal{P}_2 \cup e_2$ to $\mathcal{T}$;
17   **end**
18   Refresh $\underline{\underline{A}}^v$;
19   **end**
20   **end**
21   **end**
22   **end**
23 **end**

6.4 G-NFL Heuristic

case. We note here that the Network-Wide Local Link Code Construction (NL-LCC) algorithm presented in Algorithm 5 is subject to $O(|V| \cdot |E|^2 \cdot (|E| + |V| \log |V|))$ computational complexity, as in Step (7) each edge should be considered in the graph. As $|N| \ll |E|$, the G-NFL problem is considered to be more efficient and scalable than NL-LCC.

As an example G-NFL solution obtained by Algorithm 6 is presented in Fig. 6.1.

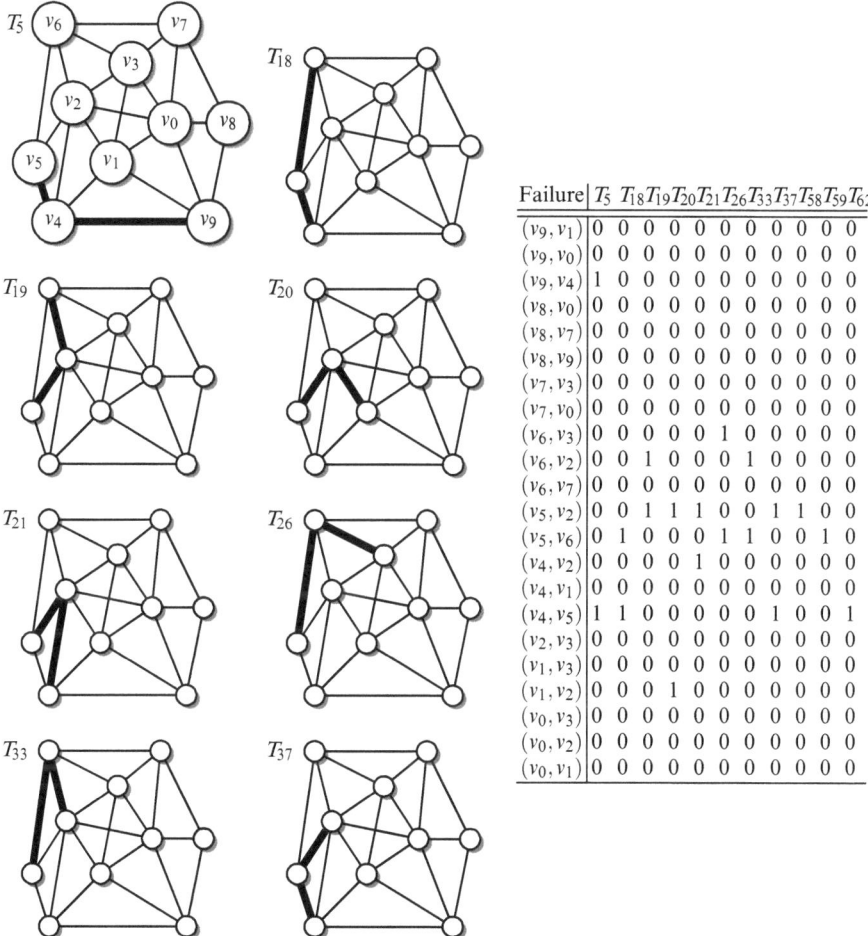

Fig. 6.1 A G-NFL solution at v_5 based on bm-trails assuming a W-LP between each node-pair. The solution has $b = 46$ and $\|\mathscr{T}\| = 92$ in addition to the 22 link-based monitors (T_{47}–T_{68}) added as an initial guess in Step (2) of Algorithm 6

6.5 Performance Evaluation

A set of experiments was conducted to verify the G-NFL scenario. Two classes of random planar graphs were generated: one for dense and the other for sparse networks, with a maximum inner face size of 4 and 7, and an average nodal degree 4.0 and 2.8, respectively. Thirty percent of all node pairs are randomly selected for being loaded, where a pair of W-LPs are shortest-path routed for each loaded node pair on the same route in both directions, which is protected by a set of P-LPs shortest and diversely routed from each link of the W-LP.

6.5.1 Size of Neighborhood

We first observe the size of neighborhood under SOD and LOD as shown in Fig. 6.2 based on the randomly selected 30% and 100% of node pairs loaded with W-LPs and P-LPs, respectively. It clearly shows that the sizes of nodal neighborhoods grow very mildly as the network size increases, compared with NL-UFL where all the links are contained in the neighborhood of each node.

We note here that the effectiveness of the FDP on NL-UFL based on NL-LCC was demonstrated and compared to traditional protection approaches, such as p-Cycles [9]. Under our G-NFL framework—as shown in Fig. 6.2—only a fraction of the links needs to be localized, i.e., based on the W-LPs and P-LPs of the same FDP solution, G-NFL will always outperform the NL-LCC framework in S-LP allocation. Thus, G-NFL will outperform p-Cycle and other shared protection approaches as well in bandwidth consumption, as already NL-LCC did it.

6.5.2 Restoration Time Analysis

Figure 6.3 shows the average diameter d of the neighborhood of each node under SOD and LOD, which corresponds to 30% and 100% of loaded node pairs, respectively. The diameter of the neighborhood serves as an important parameter to determine the maximum length of m-trails that is $d + 1$, which is further related to the maximum restoration time. Note that the restoration time of a W-LP under G-NFL can be simply modeled as the light propagation delay of the m-trails plus the latencies for LOL detection by the lambda monitors ($\sim$5 ms), nodal processing for look-up-take ($\sim$5 ms), and OXC configuration (<20 ms). Thus, for an m-trail of 2000–4000 km in length which is subject to a propagation delay of about 15 ms, we claim that the restoration time of any W-LP can be well below 50 ms.

6.5 Performance Evaluation

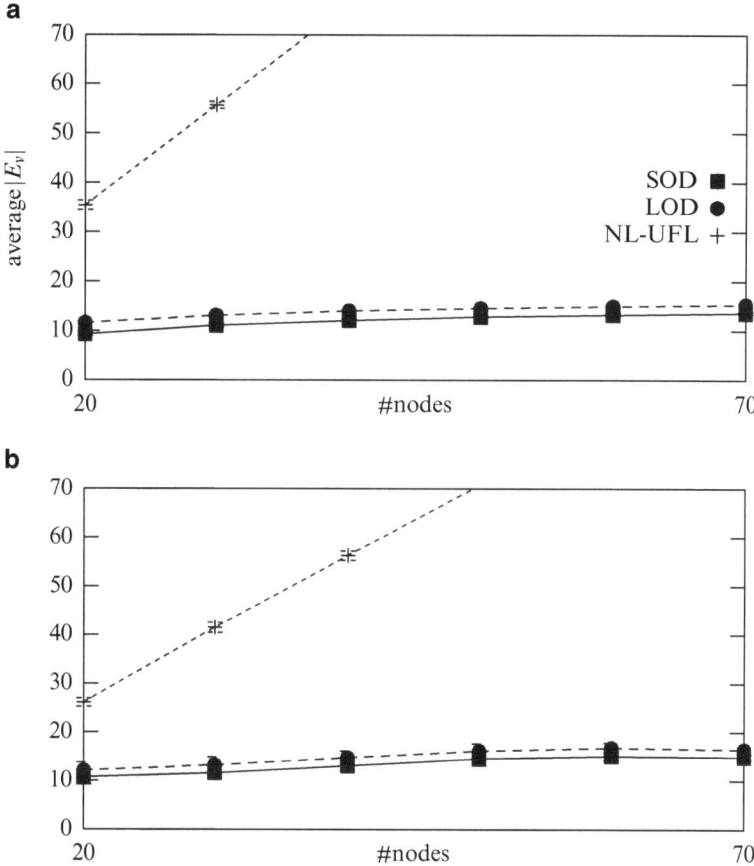

Fig. 6.2 The average size of neighborhood of each node. (**a**) Dense networks. (**b**) Sparse networks

6.5.3 Coverlength of the G-NFL Solution with FDP

Figure 6.4 shows the average WLs per link under the four definitions of nodal neighborhood. Firstly we have seen that the number of WLs per link under all the nodal neighborhood definitions scales very well when the network sizes increase; and LOD-O is outperformed by all the other schemes. This is due to the fact that it purely relies on out-of-band monitoring while under the largest possible neighborhood by most W-LPs and P-LPs (i.e., full load). The worst performance is the price paid for the complete independence between the monitoring and data planes (see Table 6.1 for details). On the other hand, SOD-IO consumes the fewest WLs per link since it jointly considers in-band and out-of-band monitoring on the smallest possible neighborhoods according to the actual W-LPs and P-LPs (i.e., 30 % loaded node pairs). The good performance is at the expense of highest m-trail reconfiguration complexity.

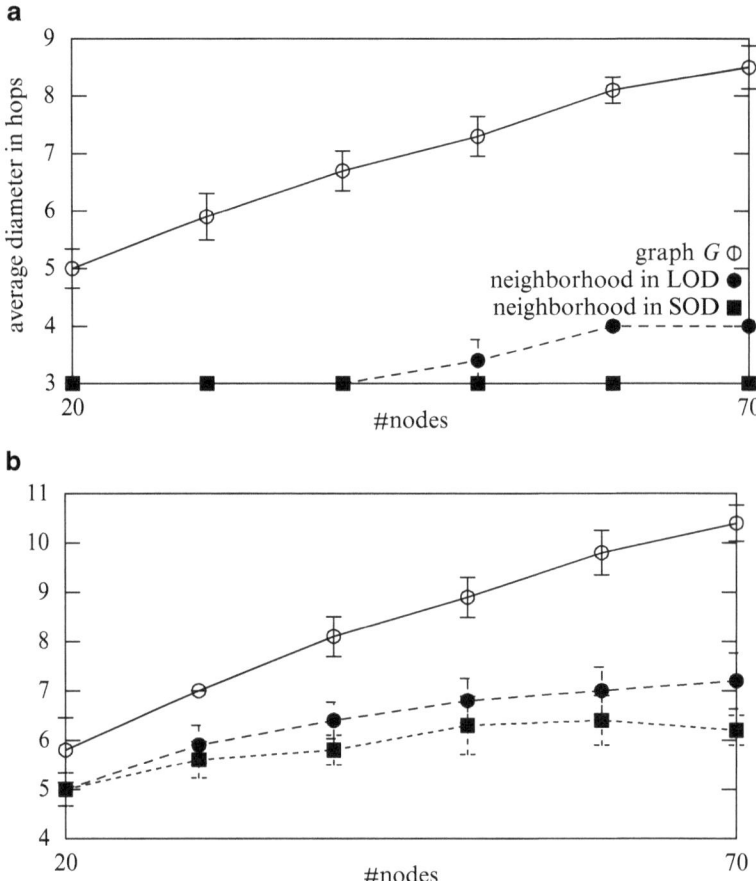

Fig. 6.3 The diameter in hops of the graph G and the neighborhood regions. (**a**) Dense networks. (**b**) Sparse networks

LOD-IO yields the second best performance among the four, and is considered a good compromise between the consumed WLs per link and the m-trail reconfiguration complexity. Note that the adoption of in-band monitoring improves performance but with little price paid as explained in Sect. 6.2.4. The results derived in Theorem 6.2 are also sketched. It is seen that some gaps exist between the derived lower bounds and each scheme, mostly due to the fact that the analysis was purely conducted based on CGT theory and can only modestly capture the additional complexity of the G-NFL problem. However, we claim that the analytical results not only contribute to the general CGT topics, but also serve as a design guideline for the G-UFL problem.

Figure 6.5 shows the average WLs per link consumed when the m-trails are allowed to reuse the spare capacity of the deployed P-LPs. We have seen that the

6.5 Performance Evaluation

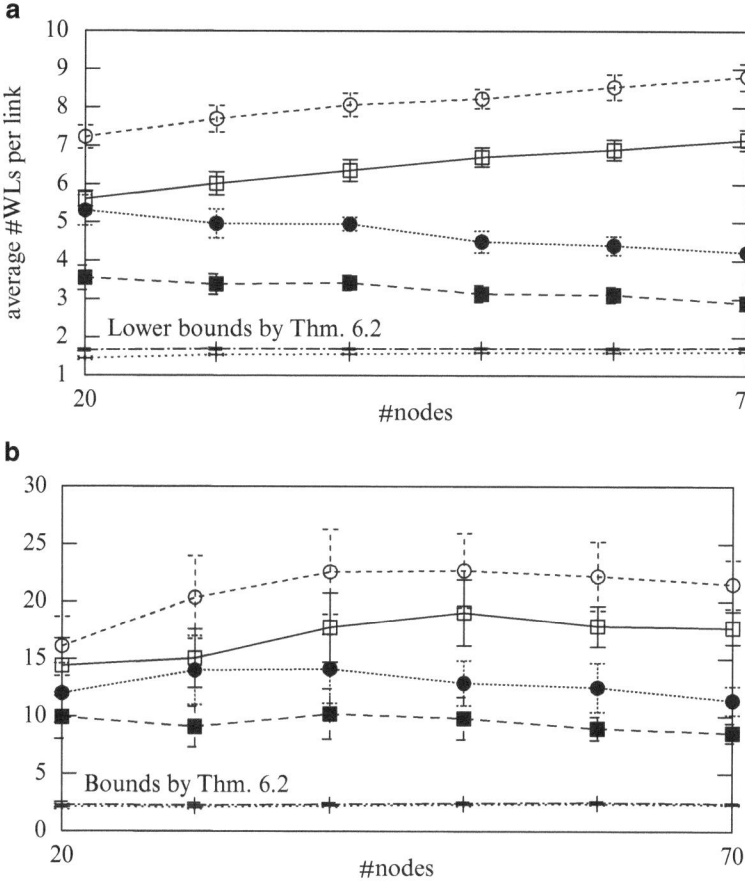

Fig. 6.4 Average number of WLs per link under the four neighborhood definitions (see Fig 6.5a for the legend) and for the lower bounds by Theorem 6.2 (denoted by +). (**a**) Dense networks. (**b**) Sparse networks

consumed WLs per link can be significantly reduced due to the reuse; and when the network sizes are increased, such monitoring overhead can be almost hidden. In this case, the network capacity efficiency approaches the efficiency of pure FDP, which has been well claimed as the optimal among all possible protection strategies.

Figure 6.6 shows the required transmitters for supporting the m-trails under each neighborhood definition. We have seen superb scalability of the G-NFL scenario, where the number of transmitters at each node, particularly for SOD-IO and LOD-IO, becomes a very small portion among the totally consumed, as the network size increases. Note that a splitter can be used at a node such that a single transmitter can support multiple m-trails.

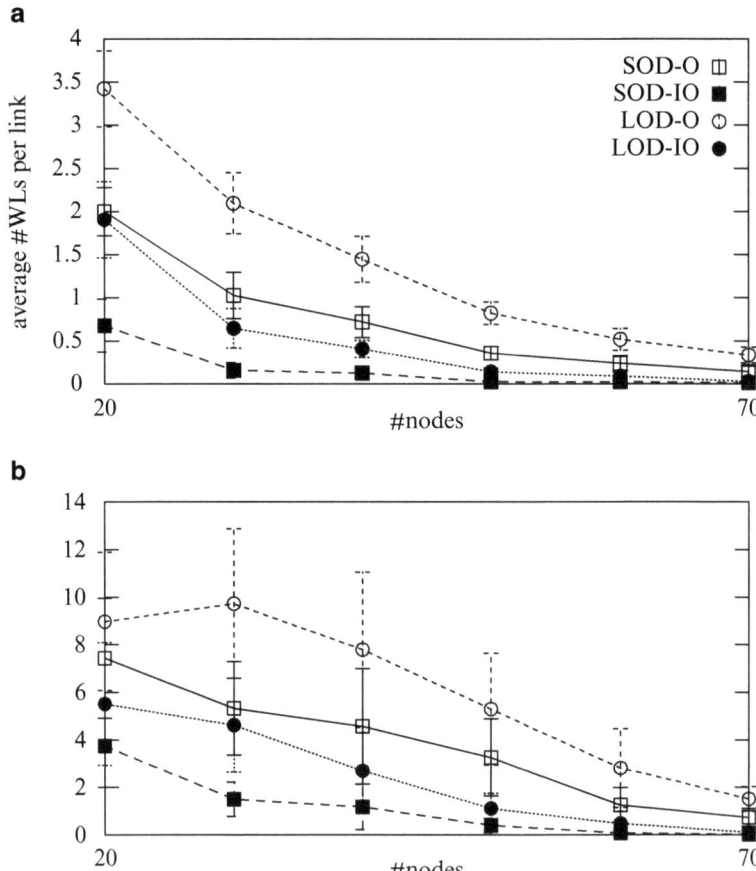

Fig. 6.5 Monitoring overhead that cannot be hidden by the spare capacity. (**a**) Dense networks. (**b**) Sparse networks

6.6 Summary

The chapter introduces an interesting scenario of m-trails, called G-NFL, which is uniquely characterized by signaling-free fault management, on-demand monitoring resource allocation, near shortest m-trails, and both out-of-band and in-band monitoring at each node. The neighborhood of a node is defined as a set of links that the node has to respond when any of them fails, by which the G-NFL problem is formulated. Bound analysis via a new CGT theory gives insight to the problem, and a simple yet effective heuristic provides a vehicle to easy solutions, which are further verified by extensive simulation results.

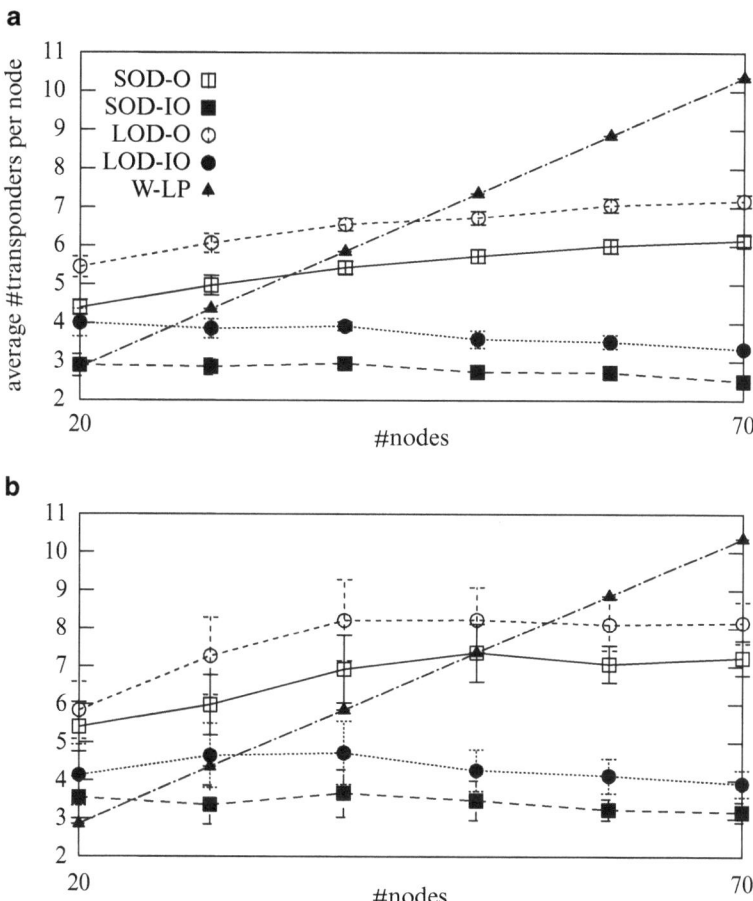

Fig. 6.6 The required transmitters in the G-NFL solution. (**a**) Dense networks. (**b**) Sparse networks

References

1. Ahuja S, Ramasubramanian S, Krunz M (2009) Single link failure detection in all-optical networks using monitoring cycles and paths. IEEE/ACM Trans Netw 17(4):1080–1093
2. Babarczi P, Tapolcai J, Ho PH (2011) Srlg failure localization with monitoring trails in all-optical mesh networks. In: Proceedings of the international workshop on design of reliable communication networks (DRCN). IEEE, Krakow, pp 188–195
3. Cisco ONS 15454 Multiservice Transport Platform (2012). http://www.cisco.com. Data sheet
4. He W, Ho PH, Wu B, Tapolcai J (2013) On identifying srlg failures in all-optical networks. Opt Switch Netw 10(1):77–88
5. Hosszu É, Moghaddam ES, Tapolcai J, Mazroa D (2013) Physical impairments of monitoring trails in all optical transparent networks. IET Netw 2(4):196–203
6. Huawei OptiX BWS 1600A DWDM System (2012). http://www.huawei.com. Product brochure

7. Tapolcai J, Wu B, Ho PH, Rónyai L (2011) A novel approach for failure localization in all-optical mesh networks. IEEE/ACM Trans Netw 19(1):275–285
8. Tapolcai J, Ho PH, Rónyai L, Wu B (2012) Network-wide local unambiguous failure localization (nwl-ufl) via monitoring trails. IEEE/ACM Trans Netw 20(6):1762–1773
9. Tapolcai J, Ho PH, Babarczi P, Rónyai L (2014) On signaling-free failure dependent restoration in all-optical mesh networks. IEEE/ACM Trans Netw 1–12. doi:10.1109/TNET.2013.2272599
10. Wu B, Ho PH, Tapolcai J, Jiang X (2010) A novel framework of fast and unambiguous link failure localization via monitoring trails. In: Proceedings of the IEEE INFOCOM WIP, San Diego, pp 1–5

Chapter 7
Dynamic Survivable Routing with M-Trails

Abstract We have seen in the previous chapters that m-trails can be incorporated with a survivable routing scheme in the optical network backbones such that the desired ultra-fast and all-optical restoration can be achieved with little additional cost. The chapter turns to the issues of dynamic survivable routing in mesh optical networks where connection requests for lightpaths arrive in the network one after the other with little knowledge of future arrivals. Firstly, we will review the spare capacity allocation problem under dedicated or shared protection schemes. A dynamic routing scheme, called *Dynamic Joint Design Heuristic (DJH)*, is introduced with its goal to allocate each request based on the failure dependent protection principle while reconfiguring the m-trails for achieving the desired ultra-fast and signaling-free failure restoration. DJH is featured as an FDP scheme that can jointly allocate a W-LP and its P-LPs to satisfy an arriving connection request, as well as the m-trails that should be newly added to the network, such that the W-LP can be restored in an all-optical fashion as in Chap. 6. By launching the m-trails possibly by reusing the spare capacity for P-LPs, the amount of WLs (wavelength channels) dedicated to the m-trails can be significantly reduced.

7.1 Spare Capacity Allocation in Dedicated and Shared Protection

7.1.1 Suurballe's Algorithm

Inherent in the restoration mechanisms of self-healing rings [5, 6, 9, 10, 15, 19–21], the dedicated protection (i.e., 1 + 1 or 1:1) provides a very fast restoration service at the expense of the fact that the ratio of redundancy (i.e., the ratio of capacity taken by P-LPs and W-LPs in the network) will go over 100 %. To implement dedicated protection in mesh optical networks for a single connection request, the physical routes for the W-LP and P-LP must be determined. It has been reported that the multi-link SRLG-disjoint diverse routing problem for dedicated protection is NP-complete [1, 11], while solving the least-cost link- or node-disjoint working and protection path-pair in a directed graph can be done by Suurballe's algorithm with polynomial computational complexity [17]. The following example demonstrates

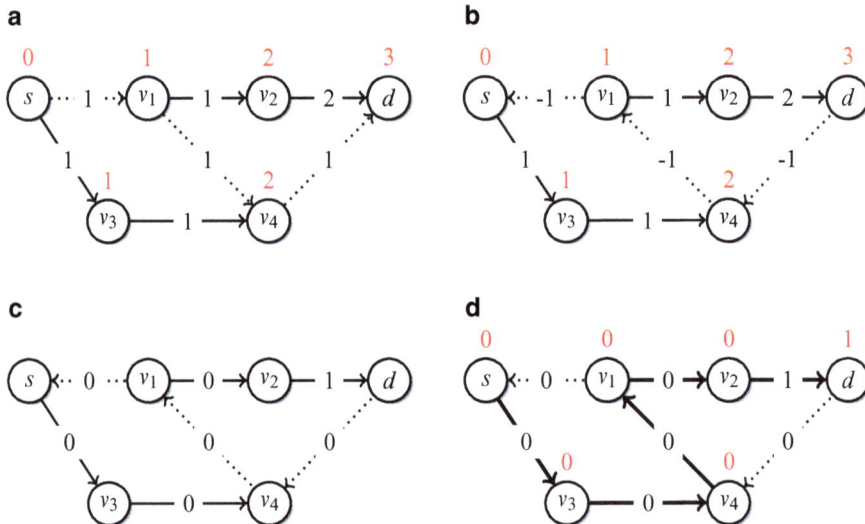

Fig. 7.1 Suurballe's algorithm (a version modified by R. Bhandari). The $d(n)$ label of node n is next to the node. The $c(i, j)$ cost of an arc is on arc (i, j). (**a**) Step 1, (**b**) Step 2, (**c**) Step 3, (**d**) Step 4

Suurballe's algorithm [16] using a 6-node network (denoted as $G = (V, E)$). The link cost is given as a label at each link, as shown in Fig. 7.1a.

Step 1 Find the shortest path *SP* from source s to destination d (e.g., $s \to v_1 \to v_4 \to d$) as shown in Fig. 7.1a. Label each node by $\pi(i)$, which is defined as the node potential determined by its cost sum from the source. The labeling of each node can be accomplished in Dijkstra's relaxation process [4].

Step 2 Reverse all arcs along the shortest path *SP*, and multiply their cost with -1, as shown in Fig. 7.1b.

Step 3 Modify the cost of all arcs according to the following formula:

$$c(i, j)' = c(i, j) + \pi(i) - \pi(j),$$

so as to get a non-negative arc cost as shown in Fig. 7.1c. The modified graph is denoted by G'.

Step 4 Find the shortest path SP' in the modified graph G', which gives SP' as $(s \to v_3 \to v_4 \to v_1 \to v_2 \to d)$ as shown in Fig. 7.1d.

Step 5 Map the shortest path of SP' to G and eliminate the common arcs of *SP* and SP' in G to obtain the *two desired paths with minimum total cost*. In the example, $s \to v_3 \to v_4 \to v_1 \to v_2 \to d$ (i.e., SP') mapped on $s \to v_1 \to v_4 \to d$ (i.e., *SP*) yields a path-pair: $s \to v_1 \to v_2 \to d$ and $s \to v_3 \to v_4 \to d$, which is nothing but the solution to the survivable routing problem.

7.1 Spare Capacity Allocation in Dedicated and Shared Protection

Suurballe's algorithm yields a computation complexity $O(|E| + |V| \cdot \log |V|)$. The term $|E|$ is due to the efforts of re-labeling all the edges in the network, and also $O(|E| + |V| \cdot \log |V|)$ is the cost of performing Dijkstra's algorithm [8]. With a joint consideration for wavelength assignment by using the wavelength graph technique [3], the problem of optimal diverse routing for dedicated protection can be solved in optical networks as well.

7.1.2 Shared Protection

Most studies on the routing of the W-LP and P-LP path-pair for shared protection are based on the technique of deriving the least-cost P-LP given the corresponding W-LP. A general method is called Two-Step-Approach (TSA) that can be applied to both shared and dedicated protection. However, the TSA could block the connection request because there is no P-LP for the given W-LP, although with the a different W-LP an appropriate P-LP could be selected. Thus, a one step approach using Suurballe's algorithm is more favorable for dedicated protection. On the other hand, for shared protection TSA is a good candidate for survivable routing.

7.1.2.1 Two-Step-Approach

TSA is an Active-Path-First scheme [22], and is the most straightforward way of solving the diverse routing problem. With the TSA, the W-LP is derived first followed by the derivation of the P-LP upon the residual network topology with the edges taken by working path excluded. If the SRLG model is considered, all the edges in the SRLGs in which the W-LP is involved should be excluded at the derivation of the P-LP. Due to the separate consideration on the two paths, the TSA algorithm can naturally solve the dependency between the working and spare capacity of each pair of links.

Let the network be denoted as $G = (V, E)$, where V is the set of nodes and E is the set of links. Let the W-LP and P-LP be denoted by W and P (the set of links the paths traverse), respectively, with bandwidth b. In the first step, W is solved with Dijkstra's algorithm using a cost function:

$$c_j^W = \begin{cases} b \cdot c_j + \epsilon, & \text{if link } j \text{ is reservable} \\ \infty, & \text{otherwise} \end{cases}, \quad (7.1)$$

where c_j is the cost for the W-LP to take a unit of bandwidth on link j, and ϵ is a small number defined as

$$\epsilon = \frac{\min_{j \in E}(c_j)}{|E|}.$$

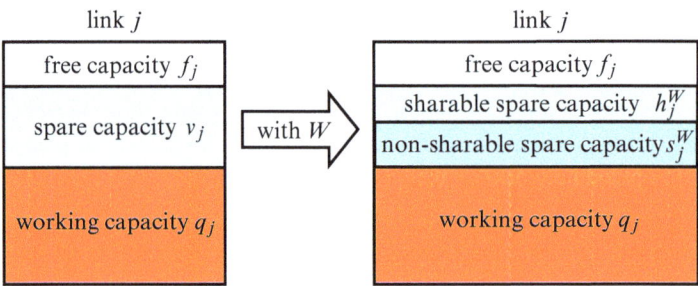

Fig. 7.2 An illustration of categories of capacity along link j

In the second step, the least-cost P-LP is solved using a different suite of cost function that considers the spare capacity sharing, discussed in the next paragraphs.

The capacity along link $j, \forall j \in E$, can be categorized into the following three types as shown in the left side of Fig. 7.2:

(1) *Free capacity* (denoted as f_j), which is the link capacity that can be reserved as either working or spare capacity.
(2) *Spare capacity* (denoted as v_j), which is the link capacity reserved by some other protection paths.
(3) *Working capacity* (denoted as q_j), which is the capacity taken by some working paths and cannot be taken for any use until the corresponding working paths are torn down.

With the presence of W, the spare capacity along link j can be further categorized into the following two types as shown in the right side of Fig. 7.2:

(1) *Sharable spare capacity* (denoted as h_j^W), which is the link capacity that has been reserved by some other backup paths, and is sharable to P.
(2) *Non-sharable spare capacity* (denoted as s_j^W), which is the link capacity that has been reserved by some other P-LPs, and is not sharable to P due to the SRLG constraint. Note that $v_j = s_j^w + h_j^w$, which is the total spare capacity along link j.

The link-state for solving the P-LP is also referred to as *spare link-state*, which must be derived before Dijkstra's algorithm is invoked in the second step. A general approach of defining the spare link-state for W is described in Eq. (7.2) [3, 14] as follows:

$$c_j^P = \begin{cases} \epsilon, & \text{if } h_j^W \geq b \\ b \cdot c_j \cdot r_j^W + \epsilon, & \text{if } h_j^W + f_j \geq b > h_j^W, \\ \infty, & \text{if } h_j^W + f_j < b \end{cases} \quad (7.2)$$

7.1 Spare Capacity Allocation in Dedicated and Shared Protection

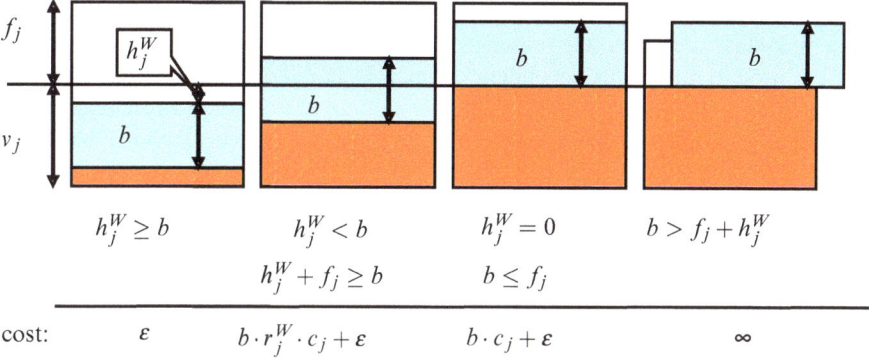

Fig. 7.3 The possible situations of a different cost function defined in Eq. (7.2)

where the possible resource sharing is

$$r_j^W = 1 - \frac{h_j^W}{b}.$$

Figure 7.3 shows the three situations defined in Eq. (7.2) for the link-states. In the left hand side scenario in Fig. 7.3, P can have all b in the sharable spare capacity region, therefore, the cost of reserving a unit of spare capacity on such a link j is ϵ, as shown in the first condition in Eq. (7.2). In the second and third scenarios in Fig. 7.3, the P-LP may partly take the free capacity region and the sharable spare capacity region; therefore, the link cost is $b \cdot c_j \cdot r_j^W + \epsilon$, which is shown in the second condition in Eq. (7.2). In the right-hand side of Fig. 7.3, the link cost is infinity because P cannot be supported by the residual capacity of the link, which is shown in the third condition in Eq. (7.2). Note that the P-LP is assumed to take sharable spare capacity along a link whenever there is any sharable spare capacity available. If there is not enough sharable spare capacity along this link to cover the total bandwidth demand for protecting W (i.e., b), P takes free capacity after considering all the sharable spare capacity.

7.1.2.2 Joint Derivation of W-LP and P-LP

For dedicated protection, the derivation of the least-cost W-LP and P-LP path-pair can be solved easily by Suurballe's algorithm. However, it is not the case for shared protection. In this section, an Integer Linear Program (ILP) is introduced for solving the shared protection problem in a single step.

As we have seen in Sect. 7.1.2.1, a proper design of spare link-state (c_j^P) is crucial to the network performance, and is still an open question subject to extensive research efforts [2,3,14,22]. The spare link-state defines the cost of the backup path of W passing through link j, in which h_j^W is the only variable that must be figured out (or equivalently, s_j^W since $v_j = s_j^W + h_j^W$). Note that h_j^W and s_j^W are network-wide link-state specific to the presence of W. The *spare provision matrix*, denoted as $\underline{\underline{S}}$, is used to model this dependency, which is an $|E| \times |SRLG|$ matrix and the entry (j, i) of $\underline{\underline{S}}$ (denoted as $s_{j,i}$), stores the amount of non-sharable spare capacity along link j for the protection path if the corresponding working path is involved in the ith SRLG entry. The most straightforward way of obtaining the matrix of $\underline{\underline{S}}$ is to simulate the failure of each SRLG and measure the amount of restoration traffic on each link. With the single failure scenario, only one SRLG could possibly be subject to an interruption at a moment. Thus, we can derive s_j^W for $f \in E$, by finding the maximum demand of spare capacity among all the SRLGs traversed by W, i.e.,

$$s_j^W = \max_{l \in W} s_{j,l}.$$

The objective of the routing problem is to minimize the total cost of the used edges in the working or any protection path, formally:

$$\min \left(\sum_{j \in W} (b \cdot c_j + \epsilon) + \sum_{j \in P} (b \cdot c_j \cdot r_j^W + \epsilon) \right). \quad (7.3)$$

However, owing to the large number of constraints, this ILP formulation leads to a tractable problem only in a small fraction of practical scenarios. Thus, an efficient heuristic approach is required, which can be easily incorporated with the S-LP allocation problem, discussed in Sect. 7.2.

7.2 Dynamic Joint Design Heuristic (DJH)

Before introducing the DJH algorithm, let's revisit an example of m-trail operation [18] discussed in Chap. 5. In Fig. 5.2 all the nodes on $P_1^{(v_3,v_4)}$ and $P_1^{(v_4,v_1)}$ can all-optically localize the failure on (v_3, v_4) and (v_4, v_1), respectively, and immediately respond to any link failure upon W_1. Similarly, all the nodes on P_2^* can all-optically localize the failure on (v_1, v_2) and immediately respond to the failure upon W_2. In other words, both W_1 and W_2 can be restored from any possible link failure in an all-optical manner. In such a situation, we define that W_1 and W_2 meet the *necessary monitoring requirement (NMR)*. It is clear that a W-LP meeting the NMR can be restored from a failure event in an all-optical and signaling-free manner; and by exercising the NMR, each network node can initiate a proper OXC configuration upon any predefined failure event merely based on the on-off status of the m-trails and W-LPs (if in-band monitoring is enabled) traversing through it.

DJH aims to obtain the W-LP, P-LPs and a set of m-trails corresponding to a connection request, such that each node along the P-LPs meets the NMR of the connection. Here, the NMR for allocating a connection refers to the situation that

7.2 Dynamic Joint Design Heuristic (DJH)

each node along the P-LPs should be able to instantly obtain the status of SRLG failures that affect the W-LP by inspecting the on-off status of the m-trails and W-LPs that traverse through the node. Then the nodes on the P-LPs can react to the identified failure immediately for the W-LP restoration without waiting for any control plane notification.

Let each connection request $C_i = (s_i, t_i)$ be bidirectional, where s_i and t_i denote the two end nodes, respectively. A WL (wavelength channel) reserved for an m-trail can be reused by a P-LP. When a new connection request C_i arrives, it is accepted only if DJH can successfully: (1) establish a W-LP for the request, (2) allocate the corresponding P-LP(s) for protecting the W-LP from any predefined failure event, and (3) reconfigure the monitoring plane (by launching some m-trails) to meet the NMR corresponding to C_i. Otherwise, the request is blocked.

Similarly when a connection C_i is torn down, some unused m-trails are torn down according to NMR condition at each node.

Algorithm 7 provides an overview on the proposed DJH algorithm, which handles each arriving request C_i in three steps, where T^I and T^O correspond to the set of in-band and out-of-band S-LPs, respectively. Firstly, procedure $WlpR$ is invoked to generate a number of k feasible W-LP candidates corresponding to the connection request C_i by using Yen's Algorithm [13] (**Step 2**). Secondly, procedure $PlpR$ is used to determine the W-LP and P-LPs corresponding to C_i, denoted as π^* (**Step 5**). Specifically, $PlpR$ identifies the SRLGs traversed by each candidate W-LP, and uses Dijkstra's algorithm to derive a set of P-LPs each

---

Algorithm 7: Dynamic Joint Design Heuristic (DJH)

Input: nt, mit
Result: Blocking Probability, Avg WL consumption
begin
1   **foreach** *connection request* $C_i = (s_i, t_i)$ **do**
2         $W \leftarrow WlpR(k, s_i, t_i)$
3   **if** $W = \emptyset$ **then**
4   block C_i
        else
5             $\pi^* \leftarrow PlpR(W, s_i, t_i)$
6   **if** $\pi^* = \emptyset$ **then**
7   block C_i
            else
8                 $(T^I, T^O) \leftarrow MtrR(\pi^*, mit)$
9   **if** $T^I \cup T^O = \emptyset$ **then**
10   block C_i
                else
11   accept C_i, apply routings for C_i
12   update WL usage
                  end
            end
        end
    end
end

disjoint from an interrupting SRLG. This comes up with a number of W-LP/P-LP(s) combinations, and the one with the highest utility, denoted as wp, is selected. The utility of a W-LP/P-LP(s) combination wp is calculated by $Utility(wp) = -NMR_r(wp) + WL_r(wp)$, where the function of NMR_r returns the ratio of *the number of SRLG pairs to be newly differentiated if wp is routed* to *the number of undifferentiated SRLG pairs before allocating wp*; and WL_r returns the ratio of *the number of free WLs after routing wp* to *the total number of WLs*. Finally, procedure $MtrR$ (**Step 8**) is used to reconfigure the m-trails for meeting the NMR of all the existing W-LPs as well as the newly allocated W-LP.

During the whole process, the request C_i is simply blocked if any of the above steps fails (**Steps 3–4, 6–7, 9–10**). Otherwise, C_i is accepted and launched in the network (**Steps 11–12**).

A unique feature of the proposed survivable routing protocol is in the task of m-trail allocation defined in $MtrR$ given in Algorithm 8. Let an *"out-of-band"* m-trail be a lightpath exclusively used for the monitoring purpose, and an *"in-band"* m-trail be an existing W-LP whose failure status is also monitored and considered in the monitoring plane. Let $|\Phi|$, $|V|$ and $|E|$ denote the number of SRLGs under consideration, the number of nodes and links of the network, respectively. All the three procedures $WlpR$, $PlpR$, and $MtrR$ are based on Dijkstra's algorithm, whose time complexity is $O(|E| + |V|\log|V|)$. The former two invoke almost a fixed number of times of Dijkstra's algorithm, while the last one determines the pairs of SRLGs that need to be differentiated at each node, and then generates a set of m-trails based on Dijkstra's algorithm by using procedure $GenMtr$. Thus, the time complexity of $MtrR$ dominates that of DJH for placing a single connection request. Hence, the time complexity of $MtrR$ or DJH is $O\left(|V|\binom{|\Phi|}{2}(|E| + |V|\log|V| + |W^c| \cdot |\Phi| \cdot |V|)\right)$.

The following two subsections explain the two procedures $MtrR$ and $GenMtr$.

7.2.1 Procedure of $MtrR$

The procedure $MtrT$ is for m-trail allocation in response to any traffic change, and its goal is to ensure the NMR for each W-LP. It is performed in the following three stages. In the first stage, the current NMR of each W-LP is evaluated (**Step 2**) according to a random sequence of W^{c1} (**Step 3**), so as to designate as few W-LPs as possible for in-band m-trails (**Step 4**). Thus, the first stage (**Steps 2–4**) takes a complexity of $O(|W^c| \cdot |\Phi| \cdot |V|)$ since at most $|W^c|$ W-LPs are examined. For evaluating whether a W-LP is taken as an in-band m-trail, at most $|V|$ ACT examinations are performed, each having a complexity of $O(|\Phi|)$.

Secondly, in case the in-band m-trails cannot fully meet the NMR of all the existing W-LPs, additional out-of-band m-trails should be launched to differentiate

---

[1]The random shuffling is with a complexity of $O(|W^c|)$.

7.2 Dynamic Joint Design Heuristic (DJH)

Algorithm 8: M-trail Routing (MtrR)

Input: WLP-PLP canidate π, tit
Result: inband m-trail set: T^I; out-of-band m-trail set: T^O
begin

1. $\quad \{T^I, T^O\} \leftarrow \emptyset$
2. $\quad$ compute NMR imposed by π and other connections in system
   $\quad$ /* deploy W-LPs as in-band m-trails   */
3. $\quad$ randomly shuffle the WLPs in system W^c
4. $\quad$ **foreach** WLP $\omega \in W^c$ **do**
   $\quad\quad |\quad T^I \leftarrow T^I \cup \omega$ if ω helps achieve NMR
   $\quad$ **end**
   $\quad$ /* pick out-of-band m-trails in system   */
5. $\quad$ **while** $T^I \cup T^O$ cannot achieve NMR **do**
6. $\quad\quad$ pick up an out-of-band m-trail ρ in system $T^O \leftarrow T^O \cup \rho$ if ρ helps achieve NMR
   $\quad$ **end**
   $\quad$ /* launch additional out-of-band m-trails   */
8. $\quad$ **while** $T^I \cup T^O$ cannot achieve NMR **do**
9. $\quad\quad$ select an undifferentiated node n_i and an undifferentiated SRLG pair $\{\phi_j, \phi_k\}$ at n_i
10. $\quad\quad \rho \leftarrow GenMtr(n_i, \phi_j, \phi_k, tit)$
11. $\quad\quad$ **if** $\rho = \emptyset$ **then**
    $\quad\quad\quad |\quad$ break
    $\quad\quad$ **else**
    $\quad\quad\quad |\quad T^O \leftarrow T^O \cup \rho$
    $\quad\quad\quad\quad$ update WL usage
    $\quad\quad$ **end**
    $\quad$ **end**
12. $\quad$ **if** $T^I \cup T^O$ cannot achieves NMR **then**
    $\quad\quad |\quad \{T^I, T^O\} \leftarrow \emptyset$
    $\quad$ **end**
    $\quad$ **return** (T^I, T^O)
end

---

some SRLGs for some nodes. Instead of directly invoking Dijkstra's algorithm to generate new routes, the procedure firstly inspects those recorded routes that have been used for launching W-LPs in the past to see if they can serve for the purpose (**Steps 5–7**). The time complexity of the second stage (**Step 5–7**) takes $O(|T^o| \cdot |\Phi| \cdot |V|)$ since at most $|T^o|$ m-trails are tested. Note that $|T^o|$ is upper bounded by $O(|V|\binom{|\Phi|}{2})$ since at most $\binom{|\Phi|}{2}$ m-trails are needed at each node for meeting NMR. Therefore, the overall complexity of Stage 2 is $O(|v|\binom{|\Phi|}{2} \cdot |\Phi| \cdot |V|)$.

Finally, as long as the current set of m-trails cannot meet NMR (**Step 8**) at n_i due to ambiguity of any SRLG pair (ϕ_j, ϕ_k) (**Step 9**), procedure $GenMtr$ is called to generate an m-trail starting from node n_i that can differentiate ϕ_j and ϕ_k (**Step 10**). If no m-trail can be generated, $MtrR$ stops; otherwise, the generated m trail is saved in T^O (the out-of-band m-trail set), where the WL usage is updated accordingly (**Step 11**). The $MtrR$ procedure goes back to **Step 8** for generating another m-trail if the NMR is still not met. In the worst case, every SRLG pair at each node must be differentiated by generating m-trails based on Dijkstra's algorithm. Thus, the time complexity of Stage 3 (**Steps 8–11**) is $O(|V|\binom{|\Phi|}{2}(|E|+|V|\log|V|))$.

From the above analysis, the time complexity of $MtrR$ is

$$O\left(|V|\binom{|\Phi|}{2}(|E|+|V|\log|V|+|W^c|\cdot|\Phi|\cdot|V|)\right).$$

Figure 7.4 gives a detailed example for the m-trail allocation process. Suppose two connections exist in the network as shown in Fig. 7.4a, b, and d. When a new connection request arrives upon v_2 and v_3, the proposed algorithm first establishes a new W-LP W_3 along with a P-LP P_3^* as shown in Fig. 7.4c. Note that W_3 is provisioned simply by using the existing lightpath of T_3, where T_3 is turned from an out-of-band m-trail into an in-band one.

To achieve all-optical restoration for W_3, all nodes on P_3^* should be aware of the link status of (v_2, v_3). The m-trail allocation procedure $MtrR$ then allocates m-trails for the nodes. At node v_4, $MtrR$ first takes W_1 as an in-band m-trail followed by taking the two existing m-trails T_1 and T_2 into its RT. Since the NMR is sufficiently met by T_1, T_2, and W_1, no additional m-trail is needed by v_4. It is also the case for v_2 and v_3, where the existing in-band and out-of-band m-trails are sufficient for the NMR in presence of the new connection. Nonetheless at v_1, just taking W_2 and T_1 cannot differentiate (v_3, v_2) from link (v_3, v_4) and (v_2, v_4). Thus to meet the NMR at v_1, a new m-trail T_4 should be generated by calling the procedure $GenMtr$.

7.2.2 Generating M-Trails by $GenMtr$

The m-trail generation procedure $GenMtr$ is detailed in Algorithm 9. Given a node v_i and a pair of ambiguous SRLGs (ϕ_j, ϕ_k), $GenMtr$ firstly collects all links with any free WL into the link set L (**Step 1**), and the links in L but not contained by both SRLGs into the link set L^{ϕ_j,ϕ_k} (**Step 2**), respectively. Then $GenMtr$ runs for at most tit iterations (**Steps 3–10**) on the pruned graph formed by the links in L^{ϕ_j,ϕ_k} to differentiate the SRLG pair at v_i.

To be specific, a link l^+ is selected randomly (**Step 4**) and a pruned graph G^* is constructed (**Step 5**). Then, the link set L_ρ, which represents the m-trail to be generated, is initialized as $\{l^+\}$ (**Step 6**). Additional links are appended to the link set L_ρ sequentially, such that as many undifferentiated SRLG pairs can be differentiated at n_i as possible (**Step 7**).

Since L_ρ may not form a valid m-trail, Dijkstra's algorithm is used to traverse through as many links in L_ρ as possible (**Steps 8–10**). If succeeds, a valid m-trail is obtained; otherwise, the current link candidate l^+ is deleted from L^{ϕ_j,ϕ_k} and the algorithm tries to generate another m-trail using the next link candidate (**Step 10**).

In **Step 8**, $twt[l']$ stores the computed weight of link l' while $wt[l']$ denotes the given link cost on l'; $GRWT$ is a constant representing the total link cost of the

7.2 Dynamic Joint Design Heuristic (DJH)

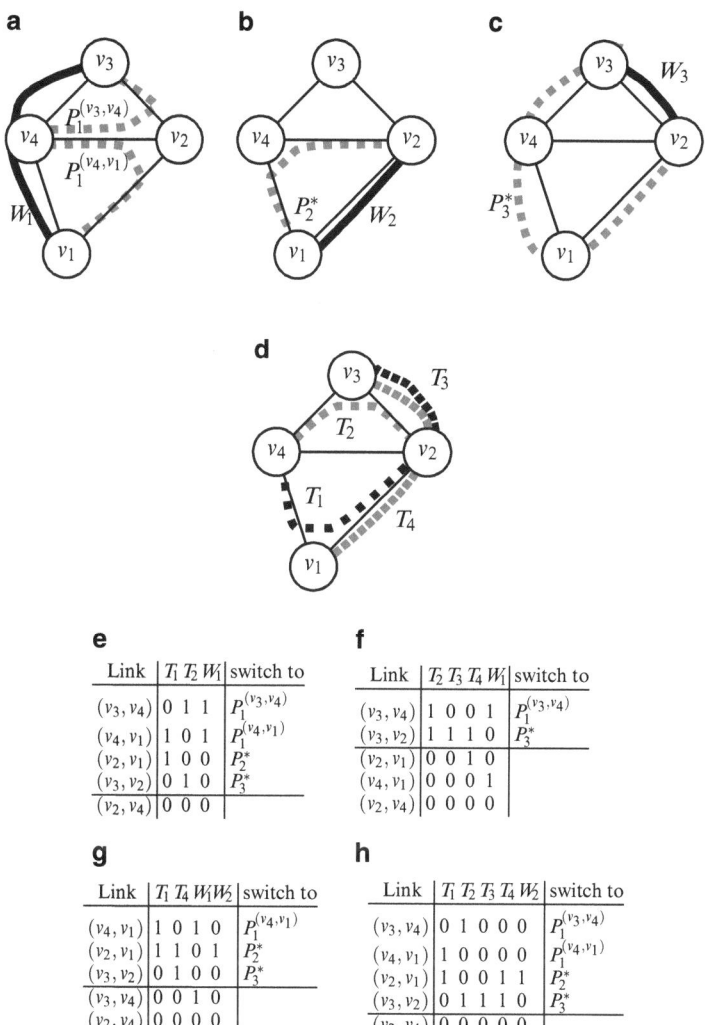

Fig. 7.4 An illustrative example of DJH extending the solution in Fig. 5.2. (**a**) Connection 1, (**b**) Connection 2, (**c**) Connection 3, (**d**) M-trails T_1, T_2, T_3, T_4, (**e**) RT at v_4, (**f**) RT at v_3, (**g**) RT at v_1, (**h**) RT at v_2

network graph while $WlUse(l')$ represents the WL usage ratio on link l'. Thus, each link in L_ρ is assigned a small weight corresponding to its given link cost while all the other links with any free WL are assigned larger weights based on their WL usage.

Figure 7.5 illustrates how T_4 in Fig. 7.4 is generated for meeting the NMR at v_1. To differentiate (v_3, v_2) from (v_3, v_4) and (v_2, v_4) at v_1, the generated m-trail should not pass through both (v_3, v_2) and (v_3, v_4) (or (v_2, v_4)). Let $GenMtr$ (as in

Algorithm 9: Generate M-trail (GenMtr)

Input: node n_i, SRLG pair: $\{\phi_j, \phi_k\}$, tit
Result: m-trail ρ to differentiate $\{\phi_j, \phi_k\}$ for n_i

1 **begin**
3     $L = \{$link with free WL$\}$; $it \leftarrow 0$
5     $L^{\phi_j,\phi_k} \leftarrow \{$link with free WL and $\in (\phi_j \setminus \phi_k) \cup (\phi_k \setminus \phi_j)\}$
7   **while** $it < tit$ and $L^{\phi_j,\phi_k} \neq \emptyset$ **do**
9        $it \leftarrow it + 1$; select a link $l^+ \in L^{\phi_j,\phi_k}$
11   **if** $l^+ \in \phi_j$ **then**
12           $G^* \leftarrow$ links with free WL and $\notin \phi_k$
13   **else**
14           $G^* \leftarrow$ links with free WL and $\notin \phi_j$
15   **end**
17        $L_\rho \leftarrow \{l^+\}$
19   **foreach** link $l' \in L \setminus \{l^+\}$ **do**
20   **if** l' helps differentiate more SRLG pairs for n_i **then**
21              $L_\rho \leftarrow L_\rho \cup \{l'\}$
22   **end**
23   **end**
25   **foreach** link $l' \in L$ **do**
26   **if** $l' \in L_\rho$ **then**
27              $twt[l'] \leftarrow \frac{wt[l']}{GRWT}$
28   **else**
29              $twt[l'] \leftarrow WlUse(l') + 1$
30   **end**
31   **end**
33        $\rho \leftarrow Dijkstra(G^*, twt, n_i,$ either end node of $l^+)$
35   **if** $\rho \neq \emptyset$ **then**
36   append l^+ to ρ if ρ does not traverse l^+
37   **break**
38   **else**
39           $L^{\phi_j,\phi_k} \leftarrow L^{\phi_j,\phi_k} \setminus \{l^+\}$
40   **end**
41   **end**
42   **return** ρ
43 **end**

Algorithm 9) choose (v_3, v_2) as the link to pass through and (v_3, v_4) as that to be disjointed. Firstly, $GenMtr$ adds (v_3, v_2) to the m-trail's link set L_ρ. Since NMR is met (by checking v_1's RT shown in Fig. 7.5b), no more link should be added. By taking each link with a unit cost, the total weight of the graph is 5. According to the weight setting rule given in $GenMtr$ (as shown in Algorithm 9: **Step 8**), the links in L_ρ are given a smaller link weight: $\frac{1}{5}$ while the other links' weights are set to 1.

Next, Dijkstra's algorithm is launched (see Fig. 7.5c) starting from v_1 to either v_3 or v_2. Let v_2 be chosen as the end node. By concatenating the obtained path with link (v_2, v_3), the m-trail $T_4: v_1 - v_2 - v_3$ is generated. After filling T_4 into v_1's RT, the NMR is met at v_1 as shown in Fig. 7.4g.

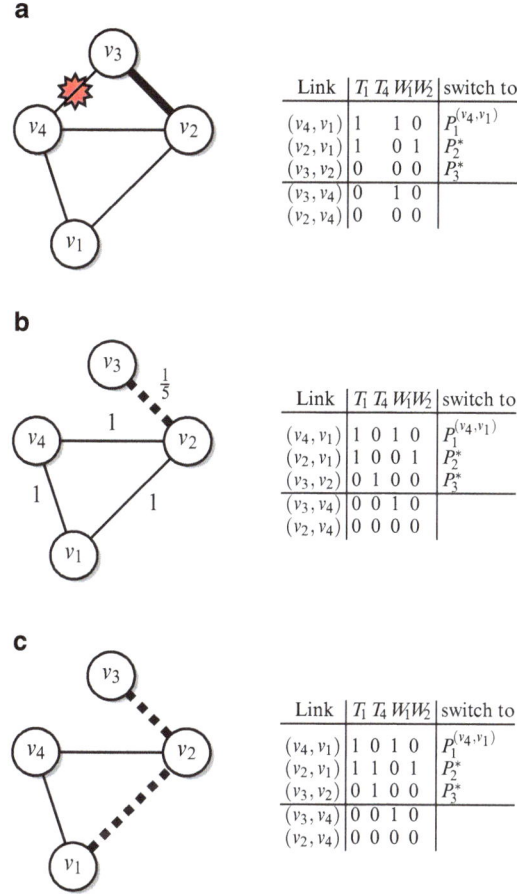

Fig. 7.5 An illustrative example on how T_4 in Fig. 7.4 is generated for meeting the NMR at v_1. (**a**) Differentiating the generated m-trail, (**b**) $GenMtr$ adds (v_3, v_2) to the m-trail, (**c**) Dijkstra's algorithm is launched

7.3 Summary

This chapter studies dynamic survivable routing in all-optical mesh networks, and introduces a survivable routing algorithm, called DJH, that jointly determines a network monitoring approach and spare capacity allocation in solving dynamic traffic requests. We find that DJH can achieve very close performance to that of DJHnt when network topologies are denser and link capacity is larger. This indicates that the deployment of m-trails in DJH would yield very little performance impact under some conditions. We believe that the proposed solution approach by incorporating the design of monitoring plane and survivable routing is a step further toward a true intelligent and autonomous all-optical networking environment.

References

1. Andersen R, Chung F, Sen A, Xue G (2004) On disjoint path pairs with wavelength continuity constraint in wdm networks. In: Proceedings of the twenty-third annual joint conference of the IEEE computer and communications societies (INFOCOM), vol 1. IEEE, Hong Kong, China
2. Bhandari R (1999) Survivable networks: algorithms for diverse routing. Kluwer, Dordecht
3. Chlamtac I, Farago A, Zhang T (1996) Lightpath (wavelength) routing in large wdm networks. IEEE J Sel Areas Commun 14(5):909–913 (1996)
4. Dijkstra E (1959) A note on two problems in connexion with graphs. Numer. Math. 1(1):269–271
5. Ellinas G, Stern TE (1996) Automatic protection switching for link failures in optical networks with bi-directional links. In: Proceedings of the global telecommunications conference (GLOBECOM), communications: the key to global prosperity, vol 1. IEEE, London, UK, pp 152–156
6. Ellinas G, Hailemariam AG, Stern TE (2000) Protection cycles in mesh wdm networks. IEEE J Sel Areas Commun 18(10):1924–1937
7. Ellinas G, Bouillet E, Ramamurthy R, Labourdette JF, Chaudhuri S, Bala K (2003) Routing and restoration architectures in mesh optical networks. Opt Netw Mag 4(1):91–106 (2003)
8. Fredman ML, Tarjan RE (1987) Fibonacci heaps and their uses in improved network optimization algorithms. J ACM 34(3):596–615
9. Grover WD, Stamatelakis D (1998) Cycle-oriented distributed preconfiguration: ring-like speed with mesh-like capacity for self-planning network restoration. In: Proceedings of the IEEE international conference on communications (ICC), vol 1. IEEE, Sydney, Australia, pp 537–543
10. Herzberg, M., Bye, S.J (1994) An optimal spare-capacity assignment model for survivable networks with hop limits. In: Proceedings of the Global Telecommunications Conference (GLOBECOM), Communications: The Global Bridge, vol 3. IEEE, pp 1601–1606
11. Hu JQ (2003) Diverse routing in optical mesh networks. IEEE Trans Commun 51(3):489–494
12. Li CL, McCormick S, Simchi-Levi D (1990) The complexity of finding two disjoint paths with min-max objective function. Discrete Appl Math 26(1):105–115
13. Martins EQV, Pascoal MMB (2003) A new implementation of Yen's ranking loopless paths algorithm. Q J Belg French Ital Oper Res Soc 1(2):121–133
14. Qiao C, Yoo M (1999) Optical burst switching (obs)–a new paradigm for an optical internet^{1}. Journal of High Speed Netw 8(1):69–84 (1999)
15. Stamatelakis D, Grover WD (2000) Network restorability design using pre-configured trees, cycles, and mixtures of pattern types. TRLabs, Edmonton, AB, Tech. Rep. TR-1999-05
16. Suurballe JW (1974) Disjoint paths in a network. Networks 4:125–145
17. Suurballe JW, Tarjan RE (1984) A quick method for finding shortest pairs of disjoint paths. Networks 14(2):325–336 (1984)
18. Tapolcai J, Ho PH, Babarczi P, Rónyai L (2013) On achieving all-optical failure restoration via monitoring trails. In: Proceedings of the IEEE INFOCOM, pp 380–384 (2013). doi:10.1109/INFCOM.2013.6566799
19. Thomassen C (1997) On the complexity of finding a minimum cycle cover of a graph. SIAM J Comput 26(3):675–677
20. Wasem OJ (1991) An algorithm for designing rings for survivable fiber networks. IEEE Trans Reliab 40(4):428–432
21. Wasem OJ (1991) Optimal topologies for survivable fiber optic networks using sonet self-healing rings. In: Proceedings of the global telecommunications conference (GLOBECOM), countdown to the new millennium. Featuring a mini-theme on: personal communications services. IEEE, pp 2032–2038

22. Xu D, Qiao C, Xiong Y (2002) An ultra-fast shared path protection scheme-distributed partial information management, part ii. In: Proceedings of the 10th IEEE international conference on network protocols. IEEE, pp 344–353
23. Xu D, Chen Y, Xiong Y, Qiao C, He X (2006) On the complexity of and algorithms for finding the shortest path with a disjoint counterpart. IEEE/ACM Trans Netw 14(1):147–158

Index

Symbols
1 + 1 protection, 17

A
AFL - Adjacent Link Failure Localization, 85, 86
alarm code table (ACT), 41

B
BER - Bit Error Rate, 10, 72
bidirectional m-trail (bm-trail), 38, 39

C
CA - Cycle Accumulation, 105
CGT - Combinatorial Group Testing, 70
chocolate bar graph, 53
circulant graph, 61
communication network, 5

D
dedicated protection, 17
dense SRLG (multiple failure), 37
device configuration (t_d), 11, 16, 152
diversity coding, 19
DJH - Dynamic Joint Design Heuristic, 187, 192
DoS - Denial-of-Service, 7
dynamic routing, 5

E
EDFA - Erbium Doped Fiber Amplifier, 10

F
failure correlation (t_c), 11, 16, 152
failure detection, 10
failure graph, 5
failure localization (t_l), 10, 16, 152
failure notification (t_n), 11, 16, 152
fault management, 10
fault restoration, 11
FDP - Failure Dependent Protection, 4, 26

G
G-NFL - Global Neighborhood Failure Localization, 39
GCS - Greedy Code Swapping, 74, 85, 94
general dedicated protection, 19
GLS - Greedy Link Swapping, 118, 135, 136, 161
GMPLS - Generalized Multi-Protocol Label Switching, 4, 9

H
HS - Hitting Set, 72

I
ILP - Integer Linear Program, 21, 22, 73, 157
in-band monitoring, 72
instantaneous recovery, 19

L

L-UFL - Local UFL, 38, 118
LCC - Link Code Construction, 84, 89, 160
LEMON - Library for Efficient Modeling and Optimization in Networks, 5
link-based monitoring, 36
LMP - Link Management Protocol, 10
LOL - Loss Of Light, 10
LOS - Loss Of Signal, 10

M

MAP - M-trail Allocation Problem, 71, 72
ML- monitoring location, 38, 71, 94, 117, 118, 152
monitoring cost (b), 40
monitoring cycle (m-cycle), 38, 39, 42
monitoring overhead, 157
monitoring resource hidden property, 157
monitoring trail (m-trail), 38, 39

N

network coding, 19
NL-UFL - Network-wide Local UFL, 39, 119, 152
NMR - Necessary Monitoring Requirement, 192

O

O/E/O - Optical-Electronic-Optical conversion, 10
OSPF - Open Shortest Path First, 9, 12
out-of-band monitoring, 72
OXC - optical cross connect, 10

P

P-LP - protection lightpath, 3, 151
path selection (t_p), 11, 16

Q

QoS - Quality of Service, 3

R

RCA - Random Code Assignment, 64
RCS - Random Code Swapping, 64, 66, 85
recovery time, 3, 15
redundancy, 11, 17
ROADM - Reconfigurable Optical Add-Drop Multiplexer, 10
RSVP - Resource Reservation Protocol, 9
RT - restoration table, 154

S

S-LP - supervisory lightpath, 4
SBPP - shared backup path protection, 4, 22
SCA - Spare Capacity Allocation, 155
shape constraint, 39
shared protection, 20, 151
SLA - Service Level Agreement, 3, 17
SLC - strong locality constraint, 66
SLP - Shared Link Protection, 24
spare capacity, 151, 156
sparse SRLG, 37
SRLG - Shared Risk Link Group, 8, 37
SSP - Shared Segment Protection, 4, 23
SSR - Successive Survivable Routing, 157
stub release, 4, 156
SURE - Strong Unambiguity RulE, 84
Suurballe's Algorithm, 187

T

TCP - Transmission Control Protocol, 12
TSA - Two-Step-Approach, 189

U

UFL - Unambiguous Failure Localization, 38, 40

W

W-LP - working lightpath, 3, 151
WL - wavelength channel, 4
WL - wavelength channels/links, 151, 156, 187

MIX
Papier aus verantwortungsvollen Quellen
Paper from responsible sources
FSC® C105338

If you have any concerns about our products,
you can contact us on
ProductSafety@springernature.com

In case Publisher is established outside the EU,
the EU authorized representative is:
**Springer Nature Customer Service Center GmbH
Europaplatz 3, 69115 Heidelberg, Germany**

Printed by Libri Plureos GmbH
in Hamburg, Germany